THE SCIENTIFIC ART OF LOGIC

THE SCIENTIFIC ART OF Logic

AN INTRODUCTION TO THE PRINCIPLES OF FORMAL AND MATERIAL LOGIC

EDWARD D. SIMMONS
Associate Professor of Philosophy
Marquette University

WIPF & STOCK · Eugene, Oregon

NIHIL OBSTAT:
    JOHN E. TWOMEY, S.T.L., Ph.D.
    Censor deputatus

IMPRIMATUR:
    ✠ WILLIAM E. COUSINS
    Archbishop of Milwaukee
    March 14, 1961

Wipf and Stock Publishers
199 W 8th Ave, Suite 3
Eugene, OR 97401

The Scientific Art of Logic
An Introduction to the Principles of Formal and Material Logic
By Simmons, Edward D.
Softcover ISBN-13: 978-1-6667-4982-3
Hardcover ISBN-13: 978-1-6667-4983-0
eBook ISBN-13: 978-1-6667-4984-7
Publication date 6/7/2022
Previously published by The Bruce Publishing Company, 1961

This edition is a scanned facsimile of the original edition published in 1961.

To my Mother and Father —
Teachers Par Excellence

Preface

This text in logic is one of the first contributions to the new Christian Culture and Philosophy Series. The book is designed generally to serve the end of the Series, and particularly to make available to undergraduate students and their instructors an elementary, but scientific, presentation of the principles of both formal and material logic. The order follows the division of logic into the logic of the first, second, and third operations of the intellect. In each section, significant logical relations, both formal and material, are examined. Since any scientific inquiry requires hard intellectual labor, it is inevitable that a scientific presentation of logic will entail some difficulty for the student. Yet this is as it must be. Formal logic is easier than material logic, but formal logic alone is an inadequate instrument of the intellect for rational discourse. Since most students take only one course in logic, it is imperative, despite the difficulties involved, that in it they be introduced to material as well as formal logic.

Since scientific inquiry is intellectually taxing, some might argue that the art of logic could be acquired much more easily in an elementary course apart from its science rather than along with it. However, the art and science of logic are indistinguishably one. Unless the rules of logical procedure are scientifically grounded in incontrovertible first principles of the logical order, they cannot adequately serve as principles either for a critique of or defense for scientific discourse. The propriety of a logical process could not be adequately defended by an appeal to a rule of logic unless that rule were itself self-evident or scientifically resolved into what is self-evident. Thus, it is an illusion to think that the art of logic could be acquired in any adequate fashion apart from the science of logic. Consequently, this text attempts to present scientifically the basic principles of both formal and material logic. Its proximate end is to generate in its users an intellectual habit which will serve as an adequate instrument for rational discourse, especially in the other sciences.

This book is not overly difficult, despite what has been said. The point rather is that it is not easy — but neither is logic. It is rigorous — so is logic. More to the point, it can be used successfully only by students who are prepared to put some effort into their work — so too logic cannot be acquired without effort. Every attempt has been made to make the presentation as straightforward as possible, given the intrinsically rigorous character of the subject matter. The opening chapter is designed, among other things, to give the student some appreciation of the nature and divisions of logic so as to orient him for the course to follow. The final chapter is devoted exclusively to the nature of logic. It is felt that at least a semester's work in logic is a prerequisite for any penetrating analysis of the nature of logic. In both the opening and final chapters there is a discussion of the division of logic into the logic of the three operations of the intellect. It is according to this division of logic that the book is divided into three parts. The opening chapter in each part includes an investigation into the nature of the intellectual operation from which that part gets its name. The remaining chapters in each part take up the logical theory pertinent to the part in question. Throughout, an effort has been made to offer sufficient examples so that the usefulness of the logical theory under discussion can be seen in a concrete setting. Each chapter is followed by exercises, which are designed to assist the student to appreciate the meaning and force of the logical theory presented in that chapter. No teacher is ever fully satisfied with another man's exercises, and every teacher has some of his own to offer to his students. However, the exercises suggested in this book are varied enough, both in format and in degree of difficulty, so that every teacher will find them to a greater or less degree of some help for his students. It is the teacher, not the textbook, who determines the program of his course. Any teacher may find that there are things treated in this book which he chooses not to include in his course. Thus, for example, a teacher might choose to pass over the discussion of the truth-functional proposition, or, perhaps, to omit the final chapter on the nature of logic. Perhaps some teacher may choose to omit some of the more difficult matter in some of the chapters. With this in mind several chapters have been ordered so that the more difficult matter is treated separately from the rest. Thus, in the chapter on relations between propositions, the case of the singular proposition — which offers many difficulties — is taken up separately. Again, since an elementary text cannot take up every question, the author has omitted any detailed discussion of the types of analogy and has discussed only the relations of the simply attributive categorical proposition in the chapter on relations between propositions. A teacher using this book may feel it necessary to supplement it with his own treatment of the division of analogy into its types and/or his own treatment of logically related modal propositions or compound propositions. Nonetheless, it has seemed to the author, after some years of experience in the teaching of elementary logic, that the subjects treated in this book are, for the most part, those which generally should and can be handled in an elementary course covering a semester's time.

Like many other textbooks, this is written within the context of the Aristotelian tradition. Thus, it bears an understandably basic resemblance to many other logic books. The rules of validity for the categorical syllogism have not changed since they were discovered by Aristotle. We are not scandalized, then, to find them repeated faithfully from textbook to textbook. They are not listed differently here. Yet this text does have several distinctively different features. As we have said, it is not limited to formal logic. Again it aims at the acquisition of the art of logic through the science of logic and not apart from the science of logic. The second part of the book includes a study of the hypothetical proposition. Among the problems taken up in this chapter are the following: Why cannot hypothetical propositions in the strict sense be truth-functional? What is the significance of a truth-functional proposition? How can symbols and even truthtables be profitably employed both for hypotheticals strictly taken and truth-functional propositions? In the logic of the third operation the chapter on the demonstrative syllogism and the consideration of self-evident propositions in the chapter on induction represent treatments necessary for an adequate course in logic which are either omitted or given scant attention in most logic textbooks. The final chapter, on the nature of logic, is an attempt to investigate this difficult subject matter on a level beyond that usually reached in logic textbooks which speak of the nature of logic only at the beginning of the book.

The book is within the Aristotelian tradition in the sense that, for primary sources, it owes most to the *Organon* of Aristotle and to logicians who have commented on the *Organon* and who have attempted to develop their own logical theory from that of Aristotle. The chapters on the categories, on the categorical proposition, on the categorical syllogism, on the demonstrative syllogism, and on fallacious argumentation owe most to Aristotle's *Categories, On Interpretation, Prior Analytics, Posterior Analytics,* and *On Sophistical Refutations.* Several chapters are indebted to Aristotle's *Topics,* especially those which treat of definition, division, and dialectical argumentation. The chapter on the predicables is first of all indebted to Porphyry's *Introduction to the Categories of Aristotle.* Other primary sources include the commentaries of St. Thomas on the *On Interpretation* and *Posterior Analytics* of Aristotle, the commentaries of both St. Albert and Cajetan on several of the logical works of Aristotle, and John of St. Thomas' *Logical Art.*

This book could not have been written without the help of many friends. They must share in whatever credit is forthcoming for the good that may be in it. The author, who alone is responsible for the shortcomings, must first of all express his gratitude to Professor Yves Simon. As a teacher beyond compare, Yves Simon has been the inspiration for this book. It most certainly will not measure up to his standards, and yet this has been the intention of the author. Many philosophy teachers, including Professor Simon, have read preliminary versions of the book. Their criticisms and suggestions have been most helpful and are most certainly appreciated. This is especially true of the men in the Department of Philosophy at Marquette University, some of whom have taught from a preliminary version of the book. The author must also express his gratitude to Marquette University for encouragement in this project and for assistance in the form of reduced schedules and secretarial help. The author is especially indebted to Mr. William May of The Bruce Publishing Company and Professor Donald Gallagher, Editor of the Christian Culture and Philosophy Series, for their editorial assistance. A special vote of thanks goes to Mr. James Campbell and Mr. Gerald Wahmhoff for service above and beyond the call of duty while serving as graduate assistants to the author during the days of composition. Finally — and most of all — the author is grateful for the patience and active assistance of his wife during the trying period of time in which this book was in preparation.

Contents

	Preface	vii
CHAPTER I.	Introduction	1
I.	The Role of Logic	1
II.	The Distinction Between Sense Knowledge and Intellectual Knowledge	2
III.	The Three Operations of the Intellect	5
IV.	Logic and Logical Relationships	7
V.	The Divisions of Logic	9
	Exercise I	11

PART I: THE LOGIC OF THE FIRST OPERATION

CHAPTER II.	Simple Apprehension: The First Operation	17
I.	The Nature of Simple Apprehension	17
II.	The Role of Abstraction in Simple Apprehension	18
III.	The Nature of Signs and Their Import in Logic	20
IV.	A Comparison of the Three Operations of the Intellect in Terms of Some Significant Distinctions	23
	Exercise II	24
CHAPTER III.	Comprehension and Extension	25
I.	The Comprehension of the Concept	25
II.	The Extension of the Concept	29
III.	The Priority of Comprehension Over Extension	30
IV.	The Inverse Ratio Between Comprehension and Extension	30
V.	Superiors and Inferiors	31
	Exercise III	33
CHAPTER IV.	The Nature of the Universal	35
I.	The Problem of the Universal	35
II.	Abstraction by Way of Simple Apprehension and Abstraction by Way of Negative Judgment	37
III.	The Division of the Universal Into the Predicables and the Categories	39
	Exercise IV	39

CHAPTER V. The Predicables — 41

I. The General Notion of the Predicables — 41
II. The Division of Universal Natures Into Five Predicables — 41
III. The First Predicable — Species — 43
IV. The Second Predicable — Genus — 44
V. The Third Predicable — Difference — 46
VI. The Fourth Predicable — Property — 47
VII. The Fifth Predicable — Accident — 49
VIII. The Limitations of This Division — 50
IX. The Tree of Porphyry — 51
Exercise V — 52

CHAPTER VI. The Categories — 54

I. The General Notion of the Categories — 54
II. Preliminary Considerations — 54
III. What Kind of Being Can Be Categorized — 57
IV. The Division of Being Into Ten Categories — 59
V. The Ten Categories Considered Singly — 62
VI. The Categories and the Predicables — 65
VII. The Modes of Opposition — 66
Exercise VI — 69

CHAPTER VII. Definition — 72

I. Definition and Division — 72
II. Nominal and Real Definitions — 73
III. The Four Causes — 75
IV. Types of Real Definition — 77
V. The Rules for Good Definition — 80
Exercise VII — 82

CHAPTER VIII. Division — 84

I. Logical Division — 84
II. Types of Logical Division — 85
III. Physical Division — 87
IV. Rules for Good Division — 88
V. Codivision and Subdivision — 89
VI. The Value of Definition and Division — 90
Exercise VIII — 92

PART II: THE LOGIC OF THE SECOND OPERATION

CHAPTER IX. Judgment: The Second Operation . . 97
    I. The Nature of the Second Operation of the Intellect 97
    II. The Pre-eminence of the Second Operation . 99
    III. Judgment and the Motive for Assent . . 99
    IV. Truth and Falsity 101
    V. The Proposition and Its Elements . . . 101
    Exercise IX . . . . , . . 104

CHAPTER X. The Supposition of Terms . . . 106
    I. Signification and Supposition . . . 106
    II. Material and Formal Supposition . . . 107
    III. The Types of Formal Supposition . . . 198
    IV. The Types of Real Supposition . . . 109
    V. Universal, Particular, and Singular Terms . . 110
    VI. Collective and Divisive Terms . . . 112
    VII. Possible and Actual Existence . . . 113
    VIII. Nonsupposing Subject-Terms . . . 114
    IX. The Unity of Terms and Logical Discourse . 115
    Exercise X 117

CHAPTER XI. The Categorical Proposition . . . 120
    I. The Difference Between the Categorical and Compound Proposition 120
    II. Simply Attributive and Modal Propositions . 121
    III. The Division of the Proposition by Way of Quality 122
    IV. The Division of the Proposition by Way of Quantity 122
    V. The Extension of the Predicate . . . 124
    VI. A, E, I, and O Propositions 126
    Exercise XI 127

CHAPTER XII. The Compound Proposition . . . 129
    I. The Nature of the Compound Proposition . . 129
    II. Conjunctive and Hypothetical Propositions . . 129
    III. The Copulative Proposition 130
    IV. Adversative and Causal Propositions . . 132
    V. The Conditional Proposition . . . 133
    VI. The Alternative Proposition . . . 135
    VII. The Disjunctive Proposition . . . 137

VIII.	Reduction of Alternative and Disjunctive Propositions to the Conditional	137
IX.	Categorical Propositions Expressed as Hypothetical	139
X.	The Symbolic Representation of Compound Propositions	139
XI.	The Truth-Functional Proposition and Its Truth Table	141
XII.	A Schematic Résumé	143
	Exercise XII	144

CHAPTER XIII. Relations Between Propositions . . 147

I.	Propositional Relations	147
II.	The Nature and Types of Opposition . .	148
III.	Subalternation	152
IV.	The Square of Opposition	154
V.	The Nature and Types of Conversion . .	155
VI.	Legitimate Conversions	157
VII.	Obversion	159
VIII.	Contraposition	161
IX.	Propositional Relations for the Singular Proposition	163
X.	Résumé	166
	Exercise XIII	167

PART III: THE LOGIC OF THE THIRD OPERATION

CHAPTER XIV. Reasoning: The Third Operation of the Intellect 173

I.	The Nature of Reasoning	173
II.	The Argument and Its Elements . .	174
III.	Validity and Truth	176
IV.	The Division of Argument . . .	178
	Exercise XIV	180

CHAPTER XV. The Nature of the Categorical Syllogism . 182

I.	The Definition of the Categorical Syllogism .	182
II.	The Elements of the Categorical Syllogism .	183
III.	Direct and Indirect Conclusions . .	184
IV.	The Figures of the Categorical Syllogism .	185
V.	The Moods of the Categorical Syllogism .	188
VI.	The Basic Principles for Categorical Syllogism .	189

VII.	An Objection to the Categorical Syllogism	191
VIII.	The Expository "Syllogism"	193
	Exercise XV	194

CHAPTER XVI. The Rules for the Categorical Syllogism . **196**

I.	The General Rules for Any Figure	196
II.	The Special Rules for Each Figure of the Categorical Syllogism	205
III.	The Valid Moods of the Categorical Syllogism	208
IV.	The Reduction of the Syllogism to the First Figure	213
	Exercise XVI	218

CHAPTER XVII. The Hypothetical Syllogism . **222**

I.	The Nature and Types of the Hypothetical Syllogism	222
II.	The Figures and Moods of the Hypothetical Syllogism	223
III.	The Simple Conditional Syllogism	225
IV.	The Case of the "Disguised" Categorical Major	227
V.	The Reciprocal Conditional Syllogism	229
VI.	The Inclusive Alternative Syllogism	231
VII.	The Exclusive Alternative Syllogism	232
VIII.	The Disjunctive Syllogism	234
IX.	The Reduction of Alternative and Disjunctive Syllogisms to the Conditional Syllogism	236
X.	Hypothetical Syllogisms With a Multimembered Major	237
XI.	Major Premises With Compound Components	239
	Exercise XVII	240

CHAPTER XVIII. Complex Syllogistic Patterns . **243**

I.	Complex Syllogisms	243
II.	The Abbreviated Syllogism	243
III.	The Syllogism With a Justification for Its Premises	245
IV.	The Polysyllogism	247
V.	The Sorites	248
VI.	The Dilemma	251
VII.	Combinations of Complex Syllogisms	253
	Exercise XVIII	254

CHAPTER XIX. Demonstrative and Dialectical Discourse . **258**

I.	The Syllogism Materially Considered	258

II.	Science and Demonstration	258
III.	Prescientific Knowledge	260
IV.	The Requirements for the Premises of Demonstration	261
V.	Universality, Perseity, and Convertibility	263
VI.	Types of Demonstration	264
VII.	The Middle Term in Demonstration	270
VIII.	Dialectical Argument	271
	Exercise XIX	276

CHAPTER XX. Induction 280

I.	The Insufficiency of Deduction	280
II.	The Types of Induction	281
III.	The Self-Evident Proposition	282
IV.	The Types of Self-Evident Propositions	284
V.	The Induction of Self-Evident Propositions	286
VI.	Mediate Induction	287
VII.	The Problem of the Inductive Enumeration	290
	Exercise XX	294

CHAPTER XXI. Fallacious Argumentation . . . 297

I.	The Notion of the Fallacies	297
II.	The Fallacies of Language	298
III.	Fallacies Beyond Language	300
	Exercise XXI	306

CHAPTER XXII. The Nature of Logic . . . 309

I.	The Purpose of This Chapter	309
II.	Art and Science	310
III.	Logic as an Art	312
IV.	Logic as a Science	312
V.	The Logic of the First Operation, the Logic of the Second Operation, and the Logic of the Third Operation	316
VI.	Formal and Material Logic	317
VII.	Logic as a Common Mode	319
VIII.	Doctrinal Logic and Logic in Use	320
IX.	The Question of Symbolic Logic	321
	Exercise XXII	323
	Index	325

THE SCIENTIFIC ART OF LOGIC

CHAPTER I

Introduction

I. The Role of Logic

There are several ways in which a man can come to know something he has not known before. The most obvious of these can be described as an immediate discovery based upon a direct sensory experience. Suppose, for example, that a man does not know that his new neighbor has several children. He will certainly learn this if he sees his neighbor's children for the first time. Another way he might learn this is to be informed of it by another neighbor who himself has seen the children. To come to new knowledge in this manner, that is, on the testimony of another, is to learn by way of faith. But there is still another way he might learn that his new neighbor has several children. If, for example, he sees several tricycles and bicycles of varying sizes in his neighbor's yard, he will know that in all probability his new neighbor has several children. To learn something in this fashion, that is, as following from something else which is already known, is to know by way of reasoning.

All men have a certain native ability to reason. No one would find it difficult to conclude in the situation cited above that the man's new neighbor has several children. This native ability is known as natural logic, and this suffices for most people in many instances of reasoning. Nevertheless, natural logic is more or less limited and subject to error. For example, it would ordinarily enable any man to know that the first of the following three reasoning processes is correct. It might or might not help him know that both the second and the third are incorrect.

1. Every animal is subject to death.
   But every man is an animal.
   Therefore, every man is subject to death.
2. Every animal is subject to death.
   But every man is subject to death.
   Therefore, every man is an animal.

3. Every animal is subject to death.
But some animals are not men.
Therefore, some men are not subject to death.

Natural logic may be strengthened by the art of logic. This art perfects man's intellect, enabling him to proceed easily and without error in reasoning, even if the reasoning process is complex and difficult. A man who has the art of logic knows without hesitation that the conclusions in the second and third examples given above simply do not follow. He knows this certainly, and he is able to explain the precise reason why the conclusions do not follow. On the other hand a man who has merely the native ability to reason might erroneously think that the conclusions follow. Even if he knows that they do not he is hard pressed to explain precisely the defect in the reasoning processes.

St. Thomas Aquinas has defined logic as "the art which directs the very act of reason, that is, the art through which a man may proceed with order, ease, and correctness in the act of reason itself."[1] We shall see later on that logic is both a science and an art. It is called the *rational science* because it is not only reasonable, which is true of every science, but because it bears directly upon the act of reason itself. It is called the *art of arts* because it directs the reason, which is the source of all the arts. We shall see that the art of logic is an instrument of the intellect in all its rational activities and especially in the rigorous procedures of scientific inquiry.

We have distinguished between natural logic, which is a native endowment, and the art of logic, which can be acquired. From now on we shall speak of acquired logic simply as logic, for our concern shall be with the art of logic rather than with natural logic.

II. The Distinction Between Sense Knowledge and Intellectual Knowledge

Logic is a perfection of the intellect, enabling us to proceed properly in the act of reasoning. In order to understand what this means, and especially to appreciate the limits of logic, we must distinguish between sense knowledge and intellectual knowledge. Human knowledge is complex, being both sensory and intellectual. Logic, however, is directly concerned only with intellectual knowledge. In the discussions which follow, it may sometimes be convenient to speak as though the intellect or the senses know. The student should not be misled by this. It is

---

[1] *Commentary on the Posterior Analytics of Aristotle*, Bk. I, les. 1.

always the man who knows, either through his senses or through his intellect.

It is not easy to *explain* the difference between sense knowledge and intellectual knowledge. The reason, perhaps, is that any explanation must be exclusively an example of intellectual rather than sense knowledge. Yet it is relatively easy to *point out* this difference. What we see, hear, smell, taste, and feel all comes under the heading of sense knowledge; what we understand is intellectual. Through our senses we become aware of the individually concrete, phenomenal characteristics of physical things. The intellect, on the other hand, cuts through the kaleidoscopic pattern of sensible characteristics to an understanding of underlying meanings. The intellect penetrates to the natures of things, to the "whatness" or quiddity or essence. Fido, for example, appears to the senses as something black and white, noisy, smelly, darting from place to place. Intellectually Fido is understood to be a dog. Fido's color can be seen, his barking heard, his odor smelled, and his movement seen. These concrete characteristics present themselves immediately to the senses. They make their impression upon the senses only as individual, fluctuating data in a concrete pattern of phenomenal or surface qualities. Fido's nature as dog differs significantly from all his concrete characteristics which we know through our senses. It does not present itself immediately to the senses. In fact, it is not known immediately, and it cannot be grasped by any sense at all. It is known only by some suprasensory power which cuts through the data of sense knowledge to the nature underlying them. The nature which makes a dog what it is cannot be seen, or heard, or tasted, or smelled, or felt. Yet the intellect, a suprasensory power, can know the nature which underlies what can be seen, heard, tasted, smelled, and felt in what we choose to call "dogs." One point should be added lest our example prove more misleading than illuminating. Fido's actual color, the noises he makes, and the odors which are peculiarly his own are known on the sense level as concrete, phenomenal data. Yet each of the sensible modifications of Fido's intelligible nature is a meaningful aspect of a reality which can be understood on a suprasensible level. Fido's concrete coloring is grasped as a visible characteristic on the level of sense knowledge, but whiteness and blackness are intelligible objects just as surely as dogness itself.

In seeing the difference between sense and intellectual knowledge it may be helpful to distinguish between various types of objects. For every consideration there is an object toward which the mind is directed. No man can totally exhaust any object in one consideration. Accordingly, we must distinguish between the thing being considered and that particular aspect of it properly and precisely known in the consideration in question. The thing in general is called the material object, whereas the precise aspect under which it is known is spoken of as the formal object. Take our knowledge of Fido, the dog. Here Fido is the material object both of the act of seeing his color and of the act of understanding or thinking his nature. Fido's color is the formal object of the act of seeing, and his nature or essence as dog is the formal object of the act of understanding or thinking. The basic difference between sense knowledge and intellectual knowledge is a difference in formal object. Sense knowledge bears precisely on the phenomenal aspect of reality, and intellectual knowledge bears precisely on the meaningful aspect of things.

One important difference between sense knowledge and intellectual knowledge is this. The object of sense knowledge is singular or particular whereas that of intellectual knowledge is, for man, characteristically universal. An object is said to be particular in this context inasmuch as it is incommunicably limited to one individual, and, in fact, to that individual in a given place at a given time. What I sense in anything belongs to that thing only in the place in which it is while being sensed and only at that time at which it is being sensed. A snapshot representing some person is adequately identified only in terms of a given individual in a given place at a given time. This is why we identify snapshots by name, date, and place. An object is said to be universal, on the other hand, inasmuch as it is applicable to many and is free from the restrictions of time and place. The nature or meaning *man* is applicable to an individual man in this place at this time. But it is also applicable to other men and to this man and other men in any place at any time. Thus what we grasp intellectually when we know a nature, essence, or meaning differs radically from which we know in an act of sense knowledge. The intelligible object is universal, communicable, and unchanging whereas the object of the senses is singular, incommunicable, and in a state of flux.

Although the objects of the senses and the intellect are radically different from one another, they do not exist in separation from one another. The sensible and the intelligible are differing aspects of the same corporeal thing. The intelligible aspect is somehow shrouded by the sensible, and we can get to know it only by first knowing the sensible and by intellectually discerning it in the sensible, as it were. The intellectual act of mentally separating the intelligible from the sensible is called abstraction. We shall say more about abstraction in the next chapter. For now, one further point remains to be noted. The intellect makes its object universal since it knows its object by way of abstraction. However, the point of departure for abstraction is the singular, and the intellect is able to reflect upon its act of abstracting and through this to the singular itself. Thus, the object of the intellect remains directly and characteristically universal, but the intellect can know the singular in an indirect and reflexive way.

III. The Three Operations of the Intellect

We distinguished between sense knowledge and intellectual knowledge in the preceding section in order to clarify the notion of logic. Logic is an art which perfects the intellect. Logic does not perfect the senses. As a matter of fact the senses cannot be intrinsically perfected because they do not need to be perfected. The senses function as well as they can apart from any acquired determination. However, the intellect is capable of further, acquired perfection, as is shown by the fact that we can distinguish between natural and acquired logic. We shall see in this section that there are three different operations of the intellect. Logic perfects the intellect in a sense for all three.

We have said that logic is a perfection of the intellect which strengthens it for the act of reasoning. But reasoning presupposes two prior intellectual operations, namely, judgment and simple apprehension. Before a man can reason he must judge, and before he can judge he must apprehend. The intellect must be well ordered on the levels of apprehension and judgment before it can be well ordered on the level of reasoning. Hence, if logic is to perfect the intellect for the act of reasoning — and it does — it must also perfect the intellect for apprehension and for judgment to the extent to which this is possible and to the extent to which this is necessary ultimately in the interests of sound reasoning. Thus, in order to understand the nature of logic and to appreciate its full extent we must consider what the three operations of the intellect are.

In the first section of this chapter we considered a reasoning process which is clearly correct. Let us use it here as an illustration of the three operations of the intellect.

> Every animal is subject to death.
> But every man is an animal.
> Therefore, every man is subject to death.

Before we can conclude, in this argument, that every man is subject to death, we must already have seen that every animal is subject to death and that every man is an animal. And before we can affirm, on the one hand, that every animal is subject to death and, on the other, that every man is an animal, we must first know the meanings of animal, subject to death, and man. Thus, reasoning presupposes a prior operation, called judgment, and judgment in turn presupposes the operation called simple apprehension.

Simple Apprehension

Simple apprehension is an intellectual act whereby we understand an object without affirming or denying anything about it. The object of simple apprehension can be referred to as an essence, a quiddity, a nature, or a meaning, so long as these terms are understood to signify any intelligible object. When I think *man* in an act of simple apprehension, I do no more than think this meaning. I neither assert that man is, nor that man is not, nor that man is such and such or not such and such. Obviously, this is a truncated type of knowing. It may be that we are only seldom psychologically aware of such a moment in knowledge. Yet, as we have seen, this is an operation of the intellect necessarily prior to both judgment and reasoning.

Judgment

Judgment is the act of the intellect affirming or denying one thing of another. It is also called the act of composition or division, because one thing is united with another in the act of affirmation or separated from the other in the act of denial. To think *man* is to apprehend, and to think *subject to death* is to apprehend. But to think *Man is subject to death* is to unite one object to another in an affirmative judgment. In similar fashion, to think *coal* is to apprehend, and to think *white* is to apprehend; but to think *Coal is not white* is to separate objects in a negative judgment. By comparison to the act of simple apprehension, judgment is a relatively complete act of knowing. In judging, the intellect takes a stand on the way things are, so that a judgment is either true or false. In simple apprehension, the intellect simply registers an intelligible aspect of things without making any assertions about it. Neither *man* nor *subject to death* is true or false. However, *Man is subject to death* is true, and *Man is not subject to death* is false.

Reasoning

Reasoning is the act of the intellect proceeding to new knowledge by way of something already known. Some judgments are so related to one another that together they give rise to a new judgment which is different from either of them. *Every animal is subject to death* and *Every man is an animal* taken together yield the conclusion *Every man is subject to* death. The judgment which is the conclusion in this reasoning process is different from either of the two judgments from which it follows. Yet we cannot simultaneously understand these two judgments without understanding the conclusion which follows from them. Although the conclusion is actually different from either of them, it is potentially contained in them. In the act of reasoning the intellect moves from previous knowledge to new knowledge potentially contained in the old. This movement of the mind from the known to the unknown is spoken of as rational discourse.

IV. Logic and Logical Relationships

Logic is the art which perfects the intellect for the act of reasoning. Reasoning is an operation in which the intellect orders the objects it knows in such a way as to proceed in an orderly fashion from old knowledge to whatever new knowledge is contained in the old. Because of the nature of the intellect and the reality it is suited to know, there are determinate rules which govern the act of reasoning. These rules are neither obvious nor self-evident. They can be acquired only with diligent study. One who knows what these rules are and why they are what they are, and only he, can reason easily and without error no matter how complex the reasoning called for. Only he can adequately defend himself when he reasons, and only he can attempt a competent critique of the reasoning processes of others. The art of logic is the habit of mind perfecting the intellect for the art of reasoning. Thus the art of logic is acquired when we understand the rules which govern rational discourse, and this, indeed, is the purpose of the course in logic.

These rules are determined by the logical relationships existing in the mind between objects known intellectually. As a matter of fact, these rules are simply the canonical statement of the natures of these logical relations. In order to understand these rules, we must study these logical relationships. In our final chapter we shall discuss the nature of logic more completely than is possible here. There we shall speak of logical relations as second intentions, and we shall discuss logic as indivisibly the art of rational discourse and the science of second intentions. Here it is sufficient to indicate briefly by way of example the nature of logical relations and to stress their significance for discourse.

Let us return once again to the first two of the three reasoning processes we have already used for illustration:

1. Every animal is subject to death.

But every man is an animal.
Therefore, every man is subject to death.

2. Every animal is subject to death.
But every man is subject to death.
Therefore, every man is an animal.

In the first example, *animal* is used as though it had a certain relationship to *man* and *subject to death*. In the technical language of logic it is used as though it were a middle term. And in the second example, *subject to death* is used as though it were a middle term in reference to the others. As a matter of fact, *animal* is related as a middle term to the other objects, and this is one of the reasons why the first conclusion follows from its premises. But *subject to death* is not related to the others as a middle term. When it is used as though it were, correct procedure is violated and thus the second conclusion is invalid. A logician knows that the first example is a good reasoning process and that the second is defective because he knows the nature of the relationships determining the rules which govern reasoning. Knowing what a middle term is, he knows that *animal* is correctly ordered in reference to the other objects in the first example and that *subject to death* is badly ordered in reference to the other objects in the second example. He knows that the first reasoning process is correct — and he knows why it is correct. And he knows that the second is defective — and why. It is not necessary that the student understand at this time what a middle term is for this illustration to be effective. But it is necessary for him to note that the property of being a middle term is a logical relationship which can belong to one object in reference to others. If an object is actually related to others as a middle term, it can be used validly in a reasoning process. If an object is not so related, then its use as a middle term will involve error. This much is all that is necessary to indicate by way of example the significance of logical relations for rational discourse and the place of logical relations in the logic course.

The property of being a middle term is just one of the relationships belonging to the objects in our example of a correct reasoning process. We could point out many more. For example, *animal* is not only related to the other objects as middle term, it is related to *man* as predicate and also as proximate genus and to *subject to death* as subject; and *man* is related to *animal* as subject and to both of the others as minor term. However, these logical relations are as unfamiliar to the beginning student as is the property of being a middle term. Here it is sufficient to say that these properties, along with many others, have their part to play in determining the status of the reasoning process in question. These are the things we shall study in logic, namely, logical relations such as the property of being a subject, a predicate, a proximate genus, a minor term, and a middle term. They are not easily understood, partially because, unlike the objects with which we are most familiar, they exist only in the mind. Yet they must be studied, because they must be understood. It is only by understanding the logical relationships which determine the character of logical discourse that one can acquire the scientific art of logic.

V. The Divisions of Logic

Our study of logical relationships is divided from one point of view into *the logic of the first operation, the logic of the second operation,* and *the logic of the third operation*. It is divided from another point of view into *formal logic* and *material logic*.

We have seen that the third operation of the intellect presupposes the second, and that the second presupposes the first. It is not surprising, then, that there are relationships of logical importance not only on the level of reasoning itself but on the level of judgment and simple apprehension as well. Because of this, logic is conveniently divided into the logic of the first operation, the logic of the second operation, and the logic of the third operation of the intellect. Each division of logic, of course, is concerned with those logical relations proper to that level. The third operation of the mind is ordered to the second because truth is secured in the latter. Yet logic as a whole is primarily ordered to the third operation because it is principally in the act of reasoning that the intellect can go wrong and is in need of perfection.

The division into formal and material logic has the following basis. In every process of reasoning there are two aspects. The first, which is based on the form of the process, consists in the arrangement of the argument. The second, which is based on the matter, is concerned with the meaningful content of the argument. A reasoning process is said to be valid or consistent when its conclusion follows from its premises, that is, when the premises cannot be true without the conclusion being true. The question of validity or consistency is one of form alone. Regardless of the meaningful content of the reasoning process, so long as the premises are correctly disposed in reference to one another the conclusion follows and the reasoning is valid. So long as the premises are badly disposed in reference to one another the conclusion does not follow and the reasoning is invalid. Both of the following two reasoning processes are valid:

1. Every intellectual being is free.
   But every man is an intellectual being.
   Therefore, every man is free.
2. Every elephant is a horse.
   But every dog is an elephant.
   Therefore, every dog is a horse.

No statement in the second example is true, whereas every statement in the first is. Each is valid because each involves a legitimate formal pattern. As a matter of fact the two arguments involve an identical formal pattern, which might be symbolically represented as follows:

Every A is B.
But every C is A.
Therefore, every C is B.

The following reasoning process, which we have already used as an illustration in another reference, is invalid, although each of its statements is true:

Every animal is subject to death.
But every man is subject to death.
Therefore, every man is an animal.

The inconsistency of this logical pattern may be more clearly evident to a beginning student in a reasoning process with the same form but with true premises and a false conclusion, as in the following:

Every Hoosier is an American.
But every Badger is an American.
Therefore, every Badger is a Hoosier.

Validity is the end or goal of formal logic, which studies those logical relations which determine the formal status of logical discourse. A knowledge of these relations should result in the ability to reason validly. However, it is not sufficient simply to be able to reason validly, for a valid process may not be true. We should be able to reason validly with truth and, where possible, with certainty and full explanatory force. In other words, we should be able to reason not only validly but with probative force as well.

Valid reasoning processes may differ significantly in probative force because of their meaningful content. We have already noted that validity is indifferent to truth. Valid reasoning processes are possible although no statement in the process is true. Even when all the statements are true, the reasoning processes may have different probative force because of differences in matter. Consider the following valid reasoning processes:

1. Every mother loves her child.
   But Susan is a mother.
   Therefore, Susan loves her child.
2. Everyone capable of speech is rational.
   But every man is capable of speech.
   Therefore, every man is rational.
3. Everyone who is rational is capable of speech.
   But every man is rational.
   Therefore, every man is capable of speech.

The first reasoning process argues validly to a *probably* true conclusion from a premise true only for the most part. The second argues validly to a *certainly* true conclusion establishing the *fact* of rationality in man from one of its effects, namely, the capability of speech. The third argues validly to a *certainly* true conclusion from premises establishing not only the fact of the capability of speech in man but also the *explanation* of this attribute in terms of its cause, namely, rationality. The three reasoning processes are equally valid, but they differ significantly from one another in terms of probative force.

Truth, certainty, and explanatory force are the end of material logic, which deals with those logical relations determining the material, or, perhaps better, truly scientific status, of logical discourse. A knowledge of these relations, in addition to a knowledge of those which are of formal import, should equip a man to reason with total effectiveness.

The purpose of logic in general is *totally* integral reasoning. This involves soundness both on the side of form and matter. Hence we shall concern ourselves, on each succeeding level of intellectual operation, with logical relations both in formal and material logic.

Exercise I (A)

Before beginning your logic course, try the following brief test. The logical processes involved are relatively simple. Yet there is a good chance that natural logic will not be sufficient for the average student to handle the test easily and without error. If this is the case for you, clearly you can benefit by your course in logic. Time yourself. A student of average competence should be able, after completing his course in logic, to get a perfect score in no more than six or seven minutes. This preliminary test in no way indicates the *full* scope of the course to follow, but it should nevertheless serve to alert most students to their need for a reflexive study of logical processes.

In the following arguments pick out the conclusions which consistently follow from their premises and those which do not follow. For the latter give the reason why they do not follow from their premises.

1. Since Americanism is opposed to Communism and Communism is opposed to Fascism, it follows that Americanism is opposed to Fascism.
2. Since no triangle has five sides, neither can any square have five sides, for no square is a triangle.
3. No canary is four-legged, because no canary is a dog and every dog is four-legged.
4. The immaterial is inconsequential; thus God is inconsequential, since God is immaterial.
5. Since all Communists are threats to the integrity of our country, then all juvenile delinquents are Communists, because all juvenile delinquents are threats to the integrity of our country.
6. Since no rectangles are three-sided, it follows that some plane figures are not rectangles because some plane figures are three-sided.
7. All men are intelligent beings, and all intelligent beings are possessed of freedom of choice; hence, all beings possessed of freedom of choice are men.
8. One cannot burn the candle at both ends and retain one's health. Accordingly, since John has not retained his health he must have burned the candle at both ends.
9. Since no syllogisms are inductive, and some syllogisms are probable arguments, then some probable arguments are not inductive.
10. Since it is true that all Hoosiers are Americans it follows that:
    10.1. It is false that all Americans are Hoosiers.
    10.2. It is true that all who are not Americans are not Hoosiers.
    10.3. It is true that some Hoosiers are Americans.
    10.4. It is true that no Hoosiers are non-Americans.
    10.5. It is false that some Hoosiers are not Americans.

Exercise I (B)

1. Define the following significant terms: natural logic, the art of logic, sense knowledge, intellectual knowledge, material object, formal object, the singular or particular (object), the universal (object), simple apprehension, judgment, reasoning, logical relation, formal logic, validity, material logic, truth, certainty, probability, explanatory force.
2. If all men are endowed with natural logic, why is a course in logic necessary?

3. If logic is acquired at the end of the logic course, how can the course escape the charge of being illogical?
4. What is the difference between a material object and a formal object? Use this distinction to contrast sense knowledge with intellectual knowledge. Suggest several other applications of the distinction between the material object and the formal object.
5. Discuss the intimate relation between sense knowledge and intellectual knowledge in man.
6. Explain the interrelation between the three operations of the intellect. How do the three operations of the intellect serve as a principle of division for logic?
7. What is the subject matter of the science of logic? What is its end?
8. Distinguish between validity and truth.
9. Explain the difference between formal logic and material logic. Interrelate this division of logic with the division of logic determined by the three operations of the intellect.

PART I

The Logic of the First Operation

CHAPTER II

Simple Apprehension: The First Operation

I. The Nature of Simple Apprehension

Simple apprehension is that intellectual act whereby we know or understand an object but do not assert anything of it. Through this act we know an intelligible object which we can call a nature, an essence, a quiddity, or even a meaning. In simple apprehension, however, we make no assertions about this intelligible object. That is, we do not judge whether it exists or not, we do not affirm or deny of it any determinations whatsoever, and we do not affirm or deny it of anything. For example, we can apprehend what *man* is without affirming *Some man exists* or *Every man is capable of speech* or *John is a man*. In our act of apprehending *man* we think simply *man* without affirming or denying anything about this object and without affirming or denying it of anything. It is difficult to appreciate the nature of this operation, for we cannot always easily recognize it in our own personal psychological experience. We can catch ourselves thinking *John is a man* or *Some men are not tall* much more readily than we can catch ourselves thinking *man* or *tall*. However, an analysis of the operation of judging, which essentially involves either affirmation or denial, shows that we must admit that simple apprehension is a distinct and prior operation of the intellect. Surely we cannot affirm *man* of *John* or deny *tall* of *man* unless we first know *man* and *tall* through acts of simple apprehension.

In comparison to the object of a judgment, the object of simple apprehension is simple or incomplex. When we make a judgment we know one thing as affirmed or denied of another, whereas in simple apprehension we know something without making any affirmations or denials about it. Nevertheless, on its own level the object of simple apprehension can be spoken of as simple or complex. It is simple if the nature it represents is simply one, and it is complex if that nature is a composite of several natures. Both *man* and *philosopher* are objects of simple apprehension. The same is true for both *stone* and *pebble*. However, *man* differs from *philosopher*, and *stone* from *pebble*. *Man* and *stone* are objects whose meanings are realized in what is distinctly one nature, whereas *philosopher* and *pebble* are objects whose meanings are realized only in a combination of distinctly different natures. *Philosopher* is equivalent in meaning to *man* plus the quality of *philosophy*. *Pebble* is equivalent in meaning to *stone* plus *smallness*. A philosopher is a philosophical man and a pebble is a small stone. *Man* and *stone* are examples of simple or incomplex objects of simple apprehension, whereas *philosopher* and *pebble* are examples of complex objects of simple apprehension.

Philosophical man and *philosopher* are identical in meaning, just as *small stone* and *pebble* are identical in meaning. Nevertheless they differ in the manner in which they are conceived. *Philosophical man* and *small stone* are complex objects mentally expressed in a complex fashion; *philosopher* and *pebble* are these same complex objects expressed simply. The same reality is signified by both *philosophical man* and *philosopher*, and the same is true of *small stone* and *pebble*. But the mode of signifying is different in each instance. And just as complex objects can be conceived incomplexly, so incomplex objects can be conceived complexly. *Man* and *rational animal* are identically the same incomplex object as far as what is signified is concerned. However, *rational animal* is thought complexly while *man* is conceived incomplexly.

Simple apprehension is not so *simple* after all. It is "simple" apprehension only in the sense that it fastens on a nature or essence without affirming or denying anything whatsoever about that nature. The object apprehended may or may not be simple. The objects of simple apprehension can be: either (1) incomplex in themselves and incomplexly conceived (e.g., *man*), or (2) incomplex in themselves and complexly conceived (e.g., *rational animal*), or (3) complex in themselves and incomplexly conceived (e.g., *pebble*), or, finally (4) complex both in themselves and as conceived (e.g., *small stone*). *Man* is clearly only an object of simple apprehension, but so too is *honest, courageous, idealistic politician fighting for the common good against the enemies of the community despite the apathy of the citizens*.

II. The Role of Abstraction in Simple Apprehension

Physical things involve an intelligible aspect as surely as they involve a set of sensible characteristics. As a matter of fact, everything involves many different intelligible aspects, including its own basic nature plus the natures of each of its many attributes. However, as noted in the preceding chapter where we distinguished sense from intellectual knowledge, the intelligible aspects of a thing are shrouded in its sensible aspects. This is true in the sense that the intelligible aspects of anything are not immediately evident to the intellect, whereas its sensible qualities are immediately evident to the senses. However, because the sensible characteristics of anything are proportioned to its nature (e.g., snow is regularly *white*), they do, in a way, manifest that intelligible nature. Through the instrumentality of sufficient sensory experience, a man is able to penetrate to underlying intelligible natures. This intellectual penetration of the sensible is abstraction. As physical things exist outside the mind, they are immediately or actually sensible and only mediately or potentially intelligible. Abstraction makes the potentially intelligible actually intelligible so that it can be simply apprehended.

There is a double force to the abstraction which yields the object of simple apprehension. First, intelligible objects as known are not limited to one individual but are universally applicable to many. The reason is that they are abstracted from the individuating characteristics which mark off one individual from another. The object *man* is mentally separated from the individuating characteristics of any individual man. Thus *man* is equally applicable to Mary, George, and Joe. In addition, the object of simple apprehension is also abstracted from the act of existing. Take the object *man* as our example again. This object is understood apart from the act of existing which makes it actual in any particular man such as Mary, George, and Joe. There is thus a distinction between *what* a thing is, its nature or quiddity or essence, and the act of existing which makes it *be*. The first operation of the intellect bears only upon quiddity or essence. It does not bear on the existence, without which, of course, no nature *is* at all. In simple apprehension a nature is known, not as actually existing, but only as open to existence. Although the nature known in an act of simple apprehension has been abstracted from actually existing things, our knowledge is not false. For we do not think that the nature does not exist, we simply think that nature without affirming or denying its existence, just as we think it without thinking about the individuating notes under which it exists in reality. The nature *man* is found in the real order only in individual men, each with his own individuating characteristics. Again, this nature is found only with the act of existing in the real order. But *man* is a legitimate object of the intellect even when apprehended apart from individuating notes and the act of existing, for neither affects in any way the character of this nature as an object of the intellect. Because the nature known in simple apprehension (e.g., *man*) is abstracted from the act of existing as well as from individuating characteristics, it is applicable not only to many actual men (e.g., Mary, George, and Joe) but to many possible men as well (e.g., Mary's great-great-grandchild, who does not actually exist but would be *man* if he did).

III. The Nature of Signs and Their Import in Logic

In order to understand more fully the three operations of the intellect we must say something about the nature and types of signs and explain their significance for logic. The notion of a sign is familiar to us. It is something which stands for, or points to, or gives knowledge of something other than itself. Signs can be divided from one point of view into *natural* and *conventional* and from another point of view into *formal* and *instrumental* signs. The reason, as we shall see more fully, why the logician is interested in signs, is that one kind of sign brings about knowledge, whereas another communicates it.

As its name indicates, a sign is *natural* whenever the connection between it and the thing it signifies is determined by nature, i.e., when it is in the nature of something to point to something else. A common example, that of smoke as the sign of fire, serves to illustrate the character of a natural sign. A fire is not the same as the smoke that it causes, but it is the nature of fire ordinarily to produce smoke. Hence, smoke is a natural indication of the presence of a fire, which may or may not be directly observed. Smoke can be understood as giving knowledge of something other than itself by pointing to fire on the basis of the natural relationship it has to fire. Other examples include falling barometric pressure as a natural sign of stormy weather, spots as a natural sign of measles and fever as a natural sign of infection, a smile as a natural sign of pleasure, and artistic works as a natural sign of rationality.

A sign is *conventional* rather than natural whenever the connection between it and what it signifies is the result of man's arbitrary imposition. For example, a green light is a conventional sign to indicate legal passage; an "A," excellence in classroom achievement; and an upraised right thumb on the part of an umpire, an "out" in a baseball game. Each of these is a sign, because each gives knowledge of something other than itself by pointing to this something else. However, none of these is natural. The connection between them and the things they signify is not natural. In each case the sign signifies what it does only because it has been chosen to signify this and is understood to do so. A red light could signify "go" and a green light could signify "stop"; an "A" could signify unsatisfactory schoolwork and any other letter could be taken to stand for excellence. And so on for the rest of our example. Granting the arbitrary character of conventional signs, we do not want to deny that some conventional signs are naturally suited to their role. Given the explosiveness of the color red it follows that a red light is better suited to signify "stop" than a blue or yellow or green light. Since "A" is the first letter of the alphabet (though this itself is, of course, something arbitrary) "A" is better suited to indicate excellence than "B," "C," "D," or "F." Nevertheless, these remain only conventional signs. Their aptness, however, is itself a natural sign of the intelligence of the men who chose them to be used as they are conventionally used.

It is not as easy to explain the difference between formal and instrumental signs as it is to explain that between natural and conventional signs. Examples of both natural and conventional signs are easily understood and readily available, but ordinarily students have no experience of formal signs to fall back upon and can clearly understand only examples of instrumental signs. *Formal* signs signify without first being known themselves. *Instrumental* signs signify only after being known themselves. Significantly all the examples we have so far given of signs are instrumental. These are the signs with which we are most familiar. We know these best, because, by definition, these signs are *themselves known* before they give knowledge of something else. The red light, for example, must be *seen* before a motorist *knows* he must stop, and many a motorist has tried to excuse himself to a traffic officer for not stopping when he should have by pleading that he has not seen the red light.

Any sign which gives knowledge of something other than itself without first being known itself is a formal sign. This means that we come to learn of these signs only indirectly and by reflecting on our direct experience of things. Paradoxical as it might seem, human knowledge is impossible without formal signs. To see why this is so, consider once more our distinction between sense and intellectual knowledge; by way of the former we grasp the sensible characteristics of things and by way of the latter we grasp intelligible natures. Neither can be understood in terms of sheer physical possession. We do not physically possess what we know either sensibly or intellectually, but represent it to ourselves in either a sensory or intelligible way. When I sense the phenomenal characteristics of something, those characteristics are present to me in an image which enables me to know them as they exist in the thing. But *what* I sense is certainly not the image *through which* I sense, but the sensible characteristics of the thing. Experience testifies to this, yet this cannot be explained save in terms of an image which represents within me what is sensible in the thing. Since the image gives knowledge of the sensible aspect of the thing, it can be called a sign. Since it signifies without first being known itself, it is a formal sign. We do not say that the image can never be known. It can be known reflexively, just as we are knowing it at this moment. But it is not known at the moment that it functions as a sign. It is not known at the moment it presents the sensible object to the knower, by interiorly representing it. At that moment it is a pure sign with nothing more to it than the object it signifies, so that this object and nothing more is known. Just as sense knowledge requires a formal sign, called the sensory image, so intellectual knowledge requires a proportionately more excellent formal sign. The object of simple apprehension is mentally expressed in a formal sign called the mental word or formal concept; what is known on the level of judgment is mentally expressed in a formal sign called the mental proposition; and what is known on the level of reasoning is mentally expressed in a formal sign called the mental argument. In each of these cases the formal sign interiorly expresses the object of whatever cognitive operation is in question: sensing, simply apprehending, judging, or reasoning. In each case knowledge is impossible without the formal sign, but in no case is the formal sign itself directly known. Formal signs, of course, are natural and never conventional since they are, as signs, the natural representations of what they signify.

We said that signs are of interest in logic because knowledge is brought about and communicated through signs. We have seen how knowledge is brought about through formal signs. But what kind of a sign is used for the communication of knowledge? Knowledge, of course, is ordinarily communicated by way of language, and words are the fundamental elements of any language. A word, in brief, is a graphic or oral expression conventionally designed to signify some meaning. Words are signs of meanings. They are instrumental signs, since they must be seen or heard before they can function as signs. And they are conventional signs, since the relation between them and the natures they represent is arbitrary, the work of men. If, for example, I want to let someone know that I am thinking about man, I must say "man." Because others understand what "man" stands for, they will know what I am thinking. I might say "l'homme" or even "homo," and if others know French and Latin either of these words serves as well as "man" to signify the meaning man. As a matter of fact we might have devised a code in which "horse" stands for the nature man and "man" stands for the nature thunder, and so on. Words are clear-cut examples of instrumental signs which are conventional rather than natural. Thus, since knowledge is communicated by words taken either singly or in combination, it is clear that knowledge is communicated by instrumental and conventional signs.

Strictly speaking, the logician is only indirectly concerned with formal signs and words. The art of logic directs the ordering of the objects of intellectual knowledge. Hence the logician is directly interested in ordering these objects, not the formal signs through which they are known. Yet, since these objects are known and expressed only through formal signs, the former cannot be ordered properly unless the latter be likewise ordered. Nevertheless the logician is content to leave the direct investigation of formal signs to psychology or epistemology. As a matter of fact the logician's concern is not even directly with the objects signified by the formal signs. The logician does not study the objects — leaving them to the sciences concerned with real things — but the logical relations belonging to them as known. The logician is even less concerned with the oral or written expression of intelligible objects. He is not directly concerned with the way in which what is known is spoken or written, but with the way in which it is ordered within the mind. He is neither a linguist nor a grammarian. Yet he cannot escape language and grammar. The linguistic and grammatical expression of his thought reflects its interior status. He can learn from teachers only by listening to their words, and he can teach his students only by speaking to them. Hence he must be interested in the oral or written expression of the objects of intellection. Yet it is clear that the ultimate verification of his statements must be found within the realm of objects and their interrelationships and not in the sphere of grammatical expression. On this point it is enough to remark that words are *conventional* signs and that the logician seeks to know the canons of rational discourse *demanded* by the nature of the reason.

IV. A Comparison of the Three Operations of the Intellect in Terms of Some Significant Distinctions

We have distinguished between the object of the act of knowing and its two expressions, mental and verbal. The object of simple apprehension is a nature or quiddity, about which nothing is either affirmed or denied. The mental expression or formal sign of this object is called a concept, since its formation within the mind is analogous to biological conception. The nature or quiddity known in simple apprehension is the object signified by this concept. Because this object is the object-of-the-concept, it has been called the objective concept. The mental expression of the objective concept, which is the concept *in the strict sense*, is called the formal concept. The formal concept is sometimes referred to as the mental word, the mental term, or the idea. The verbal expression of what is known on the level of simple apprehension is a word or combination of words spoken of as an oral or written term. We have noted already that the formal sign for the second operation is called the mental proposition, whereas that of the third operation is called the mental argument or argumentation. The corresponding verbal expressions on the levels of judgment and reasoning are spoken of as oral or written propositions and argumentations.

Throughout this book we shall be concerned with both mental and verbal terms, propositions, and arguments. For the most part our interest will be focused upon the objects they express and the logical relations ordering these objects one to another. Thus, when we speak about terms, propositions, and arguments, we will be referring primarily to the objects of the first, second, and third operations of the intellect. For the same reason, whenever we speak of the concept without distinguishing between formal concept and objective concept we shall be speaking primarily of the latter, although the name "concept" belongs first of all to the former.

Exercise II

1. Define the following: simple apprehension, judgment, reasoning, abstraction, essence, existence, sign, natural sign, conventional sign, formal sign, instrumental sign, mental term (or formal concept), objective concept, mental proposition, mental argumentation, verbal term, verbal proposition, verbal argumentation.
2. Compare these objects of simple apprehension as to complexity and incomplexity, both in themselves and as conceived: giant, science, scientist, pretty girl, girl, horse, colt, rational animal, four-sided polygon, tall man with dark brown hair, and policeman.
3. Contrast simple apprehension with judgment in terms of the complexity of their respective objects.
4. What is the twofold force of the abstraction proper to simple apprehension? What is its significance for the universality of the objective concept?
5. What is the difference between a natural sign and a conventional sign naturally suited to be used as it is used?
6. How do formal signs differ from all other signs? What are the only examples of formal signs?
7. Classify the following as signs: ringed moon as a sign of stormy weather, "dog" as a word-sign of the nature dog, the idea dog in the mind, the sensory image of a given dog, a ruddy complexion as the sign of health, a cross as the sign of Christianity, "!" as a sign of exclamation, a camel as the trade-mark for a certain type of cigaret, "SOS" as a sign of peril, and a moan as a sign of pain.
8. In which of these three are we most interested in logic: the mental term, the verbal term, or the objective concept?

CHAPTER III

Comprehension and Extension

I. The Comprehension of the Concept

Every objective concept has both a definite intelligible content and the ability of being said of many subjects. The content of the concept can be called its signification or *comprehension*. The sum of the subjects of which it can be said can be called its *extension*. We shall see that the extension of a concept depends upon its comprehension and that these two are inversely proportioned to one another.

Two concepts can differ comprehensively in at least three different ways. First of all, they can be so radically different as to involve no common element. This is the way *man* differs from *pretty*, or *running*, or *bigger*. Later we shall see that concepts of this kind are in different categories. Second, concepts may differ in meaning while admitting a common element. This is the way *man* differs from *horse*, or from *maple tree*, or from *stone*. To be a man is not to be a horse, but both are animal. To be a man is not to be a maple tree, but both are living. And to be a man is not to be a stone, but both are corporeal. In all of these examples there is a common element which partially explains their comprehension. Concepts differing from one another in this way are sometimes said to be disparate concepts. Finally, concepts may differ when one is an element which partially explains the comprehension of the other. This is the way *man* differs from *animal*, or from *living body*, or from *body*. To be a man is to be a certain kind of animal, and to be an animal is to be a certain kind of living body, and to be a living body, of course, is to be a certain kind of body. Thus *animal* is part of the comprehension of *man*; *living body* is part of the comprehension of *animal* and *man*; and *body* is part of the comprehension of *living body*, *animal*, and *man*. Concepts which differ in this way can be said to be in the same logical line. They can be compared as the more general (*animal*) to the less general (*man*), or, perhaps better, as the indeterminate (*animal*) to the determinate (*man*).

Our consideration of the way concepts can differ helps us to see more clearly what comprehension involves. The second and third ways, in particular, point out that a concept is a total meaning and that its comprehension consists in a number of intelligibly distinct notes. Thus it follows that a concept is more or less rich in content as it involves more or less intelligible notes. Any concept can be analyzed into a number of meaningful notes. Taken together, these notes give it its meaning. For example, the concept *parallelogram* is a total meaning which includes several intelligibly distinct notes. A parallelogram is a four-sided, rectilinear, plane figure with opposite sides parallel. The concept *square* is also a total meaning with several intelligibly distinct notes. It is a richer object than *parallelogram*, for it includes all the notes of *parallelogram* and more, because it is an equilateral, rectangular parallelogram. Evidently *four-sided*, *rectilinear*, *plane figure*, etc., are intelligibly distinct notes although they need not exist in separation from one another. *Plane figure* is intelligibly distinct from *rectilinear* and *four-sided* since many plane figures (e.g., circles) are neither rectilinear nor four-sided. And *rectilinear* is distinct not only from *plane figure* but also from *four-sided*, since some rectilinear figures (e.g., triangles) are not four-sided. Nevertheless these notes are not existentially distinct in any square. This is an important point. They are completely misunderstood if they are regarded as physically distinct aspects of reality somehow joined together in things. As one geometrical entity, any square is wholly a parallelogram and wholly equilateral and rectangular. It is not first a parallelogram to which rectangularity and equal sides are added. However, we can think of what makes any parallelogram a parallelogram without thinking about what makes it a square or any other kind of parallelogram. And this is true, although there is no parallelogram which is not a square or some other specific type of parallelogram. Furthermore, whatever makes a parallelogram a square is intelligibly distinct from whatever makes it a *parallelogram*. Taken together, both these intelligibly distinct notes enter into the meaning of *square* and are called its comprehensive notes. The total signification or comprehension of any concept is, of course, the sum of these notes. They are intelligibly distinct even though they need not be physically distinct. And the distinctions we draw between them in the mind are based upon differences in things. The intelligible distinction between *plane figure* and *rectilinear* holds even though every parallelogram is both rectilinear and a plane figure. This is clear since every circle is a plane figure without being rectilinear.

Concepts which are complex in themselves, such as *pebble*, include a number of more elementary intelligible notes. The reason is that they represent a plurality of different natures. *Pebble*, for example, is a total meaning made up of the meanings *stone* and *small*. Yet even concepts incomplex in themselves involve distinct intelligible notes. Thus *man* is a total meaning in which we can discover the following comprehensive notes: *substance* (that kind of a being apt to exist in itself and not in another), *material* (having parts outside of parts), *living* (able to operate vegetatively), *sentient* (able to operate on the sensory level), and *rational* (able to reason). *Animal* is an incomplex concept in the same logical line with *man* although its content is less rich than *man*. The comprehensive notes of *animal* are *substance*, *material*, *living*, and *sentient*. *Brute* is an incomplex concept sharing a common element with *man*. Just as *man* adds *rational* to *animal*, so *brute* adds its own distinctive difference to *animal*. *Running* is a concept whose comprehension is completely different from *man*. They share no common element, not even the most general, for just as *man* is ultimately reduced to *substance*, so *running* is reduced to *action*.

It is legitimate to speak of comprehensive notes as intelligibly distinct from one another. But we must be careful not to think that each of these notes is on a par with one another. One is always more general than the other. The more general is always related to the other as matter to form, that is, as the determinable to the determining factor. The more general is open to the more precise, and the latter can be understood only in the light of the former. Thus, *substance* and *material* are two of the comprehensive notes of *man*, of *animal*, and of *living body*. Together *substance* and *material* constitute the concept *body*. However, they do not constitute the concept *body* as two parts on a par with one another. *Substance* is the more general note which is open to further determination by way of *material* (or its opposite *immaterial*). *Substance* expresses indeterminately what *body* expresses determinately, and *body* does so because it adds to *substance* the determining form or difference, *material*. *Material* is less general than *substance* since some substances are not material, and *material* presupposes *substance* since everything material is fundamentally substantial.

The comprehensive notes of a concept belong to it at least implicitly or virtually. We may or may not explicitly think about the distinct intelligible notes which belong to a concept. Yet the concept is what it is because of these comprehensive notes, and in apprehending it we grasp them. If we know them explicitly, as we do when we completely define the nature known, we are said to know them distinctly. There is the temptation to look upon simple apprehension as an operation whereby the intellect immediately and intuitively understands essences in the fullness of their intelligibility. This is not the case. Simple apprehension itself must be preceded by sufficient sensory experience to warrant the abstraction of the universal from the singular. And one act of simple apprehension differs greatly from another. The natural movement of the intellect is from the general to the particular and from the confused to the distinct. The general notion of *body* is understood long before *animal* is grasped. And *animal* is understood before *man*. Our knowledge of *man* is obscure at first, that is, we grasp its comprehensive notes only implicitly before we finally reach a complete definition. It is significant that the first operation of the intellect is perfected in defining its object. The act of defining is, in its most perfect instance, the act of making explicit the comprehensive notes of the object we are defining. We cannot perfectly define many objects because we cannot fully make explicit the notes we grasp confusedly in the object. Later we shall say that *rational animal* is a perfect definition of *man*. But we shall not be able to define *brute* perfectly. We know that, like *man*, *brute* is *animal*, and that instead of being *rational*, *brute* is something else. But what is the difference which is added to *animal* to make *brute*? We are unable to name it. Instead we use the negation of *rational* in its place, for surely, whatever it is, it is *nonrational*. *Nonrational animal* suffices to point out *brute* as distinct from *man*. However it falls short of being a perfect definition of *brute* because it does not set forth, in a positive way, the comprehensive notes of *brute*. There is an opaqueness to the human intellect, testified by our inability to give a perfect definition of many objects.

In the strict sense only those notes which are *intrinsic* to the nature or essence conceived are comprehensive notes. In a less strict sense, any attribute which flows necessarily from an essence and belongs always and exclusively to that essence can be called a comprehensive note even though it is not properly intrinsic to the essence. We shall speak about this kind of attribute later as a property. *Substance, material, living, sentient,* and *rational* are the comprehensive notes of *man* in the strict sense because they are the meaningful elements which together necessarily constitute the meaning *man*. However, because *man* is what it is, certain properties always belong to every man and only to men. For example, by reason of his essence or nature, every man is able to talk, to laugh, to fashion and use tools.[1] Since the only cause of these abilities

---

[1] Throughout this book we shall refer frequently to such abilities as the ability to talk, the ability to laugh, and the ability to use tools. These will ordinarily be understood as taken *radically*, with no concern for their actual exercise or even the immediate possibility of their actual exercise. Thus a dumb man will be thought to have the

is human nature, only men can talk, laugh, fashion and use tools. These abilities and other properly human characteristics are sometimes spoken of as comprehensive notes of the concept man. However, they are comprehensive notes only in a secondary sense. They have an intimate and necessary connection with the comprehensive notes of man in the strict sense. Yet there is a basic difference between them. A man is a man because he is rational, because he is sentient, and so on. He is a man not because he is able to talk or laugh. On the contrary, he is able to talk and laugh precisely because he is a man. Whenever we speak of comprehensive notes we shall ordinarily refer to comprehensive notes strictly taken, i.e., to those meaningful notes intrinsic to the concept which taken together necessarily give us the total meaning of the concept.

II. The Extension of the Concept

Because a concept is abstracted from individuating characteristics and the act of existing, it can be said of many. Logicians call this property the extension of the concept. Extension is the sum of those subjects (either singular or universal) in which all the comprehensive notes of a concept are realized. Thus John, Peter, Paul, Mary and all other individual men fall within the extension of man, since the comprehensive notes of man, namely, substance, material, living, sentient, and rational are realized in each of them.

The extension of a concept is not measured by adding up units or by "counting heads." There is, in fact, an infinite number of subjects within the extension of any concept, and this because the concept is abstract. It is, as we have seen, abstracted both from individuating characteristics (hence man can be said of both John and Mary) and from the act of existing (hence man can also be said of Mary's great-great-grandson, who doesn't exist but would be man if he did). There is a limit to the number of actually existing men, past, present, and future, but no limit to the number of logically possible men. The same is true of any other concept, abstracting, as it must, from actual existence. Any concept includes within its extension an infinite number of subjects of which it can be said. Nevertheless, we can still speak of one concept having greater or less extension than another concept as long as we consider extension in terms of an infinite multitude and compare one infinite multitude with another. For example, man can be said of an infinite multitude of actual and possible men, while animal can be said of an infinite multitude

---

ability to talk — in the sense that his nature is such as radically to ensure this ability — even though there are physical impediments to the exercise of this ability.

of actual and possible men and brutes. Since the subjects of which *man* can be said are only some of the subjects of which *animal* can be said, the extension of *animal* can be regarded as greater than that of *man*.

III. The Priority of Comprehension Over Extension

Although comprehension and extension are both properties of the concept, comprehension has a logical priority over extension. We can distinguish between the actual content of the universal concept and the universality which enables it to be said of many subjects. The concept is first of all a meaning constituted by the sum of its comprehensive notes. It can be said of many subjects only if its comprehensive notes are found in them. Thus extension is defined in terms of comprehension, a clear indication that the latter is of primary importance.

A different opinion on the universal is held by the nominalists. In their view the universal concept is merely a collection of those subjects which answer to the name or term in question. Nor is there any necessary reason why one term is said of the members of any given group, for there is no nature which is shared commonly by the members of the group. According to the nominalists, John and Peter and Mary are men simply because they are called men. The truth is obviously quite opposite. John and Peter and Mary are called men because they are men, i.e., because each is *substance, material, living, sentient,* and *rational*.

IV. The Inverse Ratio Between Comprehension and Extension

Comprehension and extension are readily seen to be inversely proportional. The more notes in a concept, the richer it is in meaning and the fewer the subjects of which it can be said. Correspondingly, the fewer its notes, the more deficient it is in meaning and the greater the number of subjects of which it can be said. *Square* is said of fewer geometrical entities than *parallelogram*, *parallelogram* of fewer than *quadrilateral figure*, *quadrilateral figure* of fewer than *rectilinear figure*. *Sign* is said of more things than *instrumental sign*, while *instrumental sign* is said of more things than *conventional sign*, and *conventional sign* of more things than *term*. The more determinate a concept is comprehensively the more restricted it is extensively. The comprehensive notes can be likened to clues in a detective story and the extensive subjects to suspects. The more clues that are uncovered the fewer the suspects. Each succeeding clue narrows the field of suspects accordingly. Similarly, the addition of a comprehensive note to a concept yields a new concept more determinate and rich in content than the original and correspondingly more restricted in extension. For example, add to the concept *substance* the note *material*, and the concept *body* results. *Body* is said of whatever *substance* is said of, with the exception of spiritual substances. Then add *living* to *body*, and the concept *living body* or *organism* results. *Organism* is said of whatever *body* is said of, with the exception of nonliving bodies.

The inverse ratio between comprehension and extension can be seen graphically in the following schema which completes the illustration begun in the preceding paragraph:

CONCEPT   **COMPREHENSIVE NOTES**

Man	Substance — Material — Living — Sentient — Rational	Men
Animal	Substance — Material — Living — Sentient	Brutes — Men
Organism	Substance — Material — Living	Plants — Brutes — Men
Body	Substance — Material	Nonliving bodies — Plants — Brutes — Men
Substance	Substance Spirits	— Nonliving bodies — Plants — Brutes — Men

                                                EXTENSIVE SUBJECTS

Naturally the inverse ratio between comprehension and extension is significant only in reference to concepts in the same logical line.

V. Superiors and Inferiors

Those subjects, whether universal concepts or individuals, which contain the comprehensive notes of a given concept and thus fall within its extension are called *inferiors* of that concept. The concept itself, in reference to these inferiors, is called a *superior*. Superior concepts are also called *logical wholes* or *potential wholes*. Inferiors are spoken of as *subjective parts*.

These names are given from the point of view of extension and not comprehension. As a matter of fact, in virtue of the inverse ratio between comprehension and extension, these inferiors are superior and these superiors inferior from the point of view of comprehension. From the same point of view logical or potential wholes are really parts, and subjective parts are really wholes. But from the point of view of extension the concept with fewer comprehensive notes is superior, whereas the concept with more comprehensive notes is inferior. From this vantage point the less comprehensive concept can be considered a whole containing the more comprehensive concept as a part (of its extension) and the superior is well described as a potential whole. Take *rectilinear plane* figure as an example. *Rectilinear plane figure* is superior to *triangle* and to *quadrilateral figure*, since its comprehensive notes are found in both. *Rectilinear plane figure* is actually neither *triangle* nor *quadrilateral figure*, but *rectilinear plane figure* is, as it stands, potential to both the note *three-sided* and the note *four-sided*. Hence, though actually neither, *rectilinear plane figure* is potentially both *triangle* and *quadrilateral figure*. It is aptly described as a potential whole in reference to these. The inferiors are also well described as subjective parts. Whenever superiors and inferiors are universally affirmed of one another, the inferior must be used as the subject of the proposition. I can say *Every triangle is a rectilinear plane figure*, but I cannot say *Every rectilinear plane figure is a triangle*.

Objects which are inferior in one relationship may be superior in another. Thus *man* is inferior to *animal* but superior to *Mary*, and *parallelogram* is inferior to *quadrilateral figure* but superior to *square*. This is true of all universal inferiors, for they can always be considered superior at least in reference to the singulars within their extension. However, no singular inferior can ever be a superior, since singulars as such have no extension.

There is one difficulty which should be mentioned here. We have said that the superior can be said of its inferior. Thus *man* can be said of *Mary*. *Animal*, *organism*, *body*, and *substance* can all be said of *Mary* too. There is no difficulty yet, for *man*, *animal*, *organism*, *body*, and *substance* are all in the same logical line. All these concepts are hierarchically arranged in so far as they are more or less determinate expressions of Mary's essence. However, *able to laugh* and *able to talk* can also be said of Mary, and perhaps also *pretty*, *tired*, *sitting down*, and *tapping her foot*. These can be said of Mary because these too are superiors in reference to her. Mary can be a subjective part of *pretty* just as she is a subjective part of *man*. But she is not an inferior of *pretty* in the way she is an inferior of *man*. The difficulty arises because *pretty*, *tired*, *able to talk*, etc., are completely different from *man*, *animal*, *organism*, *body*, and *substance*. They have no common element with these, and must, as we shall see, be reduced to different categories. But if there is no element common to both these groups of concepts, how can Mary be their inferior since extension depends upon comprehension? To answer this difficulty we must distinguish between the nature basic to an individual and the many individuating characteristics possessed by an individual as modifications of that basic nature. Later we shall refer to this as the difference between substance and accident. It is true to say that an individual has a nature basic to him. Thus we say *Mary is a man*. It is also true to say that in an individual that basic nature is modified by individuating characteristics. Thus we say Mary is pretty. The comprehensive notes of man are realized in Mary as constituting her very essence. The comprehensive notes of pretty are also found, but they do not pertain to her essence. They do, however, enter into the definition of one of her nonessential attributes. Mary is inferior to man by way of an essential relationship; Mary is inferior to pretty by way of a nonessential or accidental relationship. Mary is inferior to man because she is essentially man. She is inferior to pretty because one of her individuating attributes includes within it the comprehensive notes of pretty. Man is an essential superior and pretty a nonessential superior in reference to Mary. Later we shall distinguish between essential and nonessential predication. Essential superiors are predicated of, or said of, their inferiors essentially, and nonessential superiors are predicated of their inferiors nonessentially. Nonessential predication is sometimes spoken of as denominative predication.

Exercise III

1. Define: comprehension, extension, inferior (or subjective part), singular inferior, superior (or logical or potential whole).
2. What is the property of comprehension? Of extension? Which is prior to which? Explain.
3. Discuss three ways in which concepts can differ comprehensively.
4. How are comprehensive notes "added" to one another to constitute the comprehension of a concept?
5. What does it mean to say that the extension of a concept involves an infinite multitude? Why does the extension of a concept involve this infinite multitude?
6. What is the inverse ratio between comprehension and extension? Why does this inverse ratio apply?
7. Does the addition of a comprehensive note to a concept make that concept more meaningful or does it make for a new concept? Explain.
8. What is the difference between knowing an object confusedly and knowing it distinctly?
9. Explain the precise significance of the terms, "logical or potential whole" and "subjective part."
10. How does an essential inferior differ from a nonessential inferior?
11. Arrange the following sets of concepts according to decreasing superiority; explain the application of the inverse ratio in each case and name the members of the sets as superiors or inferiors of one another:
    11.1. *Substance, dog, organism, brute, spaniel.*

11.2. Athlete, pitcher, ball player, baseball player, Grover Cleveland Alexander.
11.3. Hearing, ability, cognitive faculty, sense, external sense.
11.4. Moral virtue, justice, good habit, habit, perfection.
11.5. "Horse," sign, instrumental sign, term, conventional sign.

12. Prepare sets similar to those in No. 11 explaining the same points in reference to them as you did for those in No. 11.

CHAPTER IV

The Nature of the Universal

I. The Problem of the Universal

In distinguishing between sense knowledge and intellectual knowledge we have noted that the former is of the singular whereas the latter is characteristically of the universal. We have discussed the object of simple apprehension both in terms of its intelligible content and of its universal applicability to many. The universal character of our intellectual knowledge is an incontrovertible fact of introspective experience. This fact, however, has occasioned a philosophical controversy which dates back almost to the beginning of philosophy itself. In this chapter we shall look more closely at the universal, considering both the philosophical problem which the fact of the universal sets up and the proper understanding of the universal which solves this problem. The solution to the problem of the universal requires a distinction which should help considerably to introduce the chapters which follow, on the predicables and the categories respectively.

The object of simple apprehension is characteristically universal and thus can be said of many.[1] *Man*, for instance, can be said of both Mary

---

[1] The abstraction which yields a nature which can be said of concrete subjects mentally separates that nature from its individuating characteristics without, however, positively cutting it off from its concrete subjects. The reason for this is that the nature is conceived of as a *whole* which includes, in an indeterminate manner, the concrete subjects in which it is realized and which are its subjective parts. However, the same nature can also be conceived *as a part* which positively excludes all that goes with it in any concrete subject. When it is conceived of in this second way the nature cannot be said of any concrete subject. *Man, white,* and *honest* are products of the first type of abstraction, whereas *humanity, whiteness,* and *honesty* are products of the second. We can say *Peter is honest.* We cannot say *Peter is honesty.* However, we can say *Peter has honesty,* for having honesty is equivalent to being honest. *Man, white,* and *honest* are examples of concrete concepts. *Humanity, whiteness,* and *honesty* are examples of abstract concepts. To avoid confusion we must be careful to note that both are products of an abstraction, but that concrete concepts are, in a sense, less abstract and abstract concepts more abstract. Although an abstract concept cannot be said of a concrete subject, it can be said of an abstract subject. Peter is not honesty, but one of Peter's virtues is honesty.

and Peter. Man, existing in the intellect as known in its first operation, is truly communicable, that is, it can be shared by many. It is a universal object with a determinate meaning in no sense limited to any individual.

But what of the real subjects of which man can be said, e.g., Mary and Peter? Each is a concrete individual. Each is a man only because each possesses his or her own human nature. Mary's human nature is not Peter's. Mary and Peter, as they exist independently of knowledge, are, even as men, individuated, singular, incommunicable. How then can we legitimately say man of Mary, or of Peter, or of any individual man? When I say Mary is a man, am I not identifying something universal and communicable with what is singular and incommunicable? Am I not saying that Mary, unquestionably an individual man, having her own unique human nature, is the universal man or is universally man? This seems the case, since the nature (man) which I know is universal, and it is this nature which I predicate of Mary. I seem to be saying that that which is singular and incommunicable is something which is universal and communicable.

The solution to this problem requires several distinctions. Considered absolutely, that is, in itself, a nature is neither singular nor universal, but is open to either. In itself the nature man does not include universality. If it did, it could never be realized in the individual, as an incommunicable perfection. But we know that man is realized in this way in Mary and in Peter. Nor does man exclude this relation. If so, it could never be universal, as it is in the mind as an object of simple apprehension. Considered absolutely, that is, in itself, any nature is open to the state of singularity which it enjoys outside the intellect and to the state of universality which it enjoys as an object of the intellect, but in itself it includes neither of these. Neither universality nor singularity is a comprehensive note of any concept.

We can look at the universal object from two points of view, first, as a determinate meaning with a definite intelligible content or comprehension; second, as related to the possible inferiors of which it can be said. In other words, we must carefully distinguish the nature known from the relationship of universality which relates it to many. The nature and its universality are not indistinguishably one. Universality is a special form which belongs to the nature precisely as known. It results from the abstract condition of the nature as an object of simple apprehension. The nature is spoken of as universal only because it acquires universality, which, as such, is distinct from it. To employ a distinction introduced at the end of the last chapter we might say that the object of simple apprehension is universal denominatively rather than essentially.

To object to the proposition Mary is a man on the grounds that this is to identify the universal with the singular is to ignore the distinction between the nature which is universal and the universality which makes it universal. It is true that Mary is a singular, and man a universal. However, man is universal only because the mind confers universality on it, and it is only the nature and not its universality which is said of Mary or of any other individual man. In the proposition Mary is a man all that is affirmed is that the nature man is really found in Mary. This nature is considered as a meaning explained in terms of the comprehensive notes substance, material, living, sentient, and rational. And these are real, even though universality is not.

II. Abstraction by Way of Simple Apprehension and Abstraction by Way of Negative Judgment

An object becomes universal to the extent that it is abstracted from individuating characteristics. It is universal only in the mind; it is singular in the real. Since truth is a correspondence between the mind and the real some philosophers have objected that abstraction falsifies the object. After all, man exists in the mind without any individuating characteristics, but each existing man is concretely individuated. To solve the problem of the universal presented in this way, we must introduce a distinction between two modes of abstraction. These are abstraction by way of simple apprehension and abstraction by way of negative judgment.

In general abstraction can be described as a mental separation. This, however, can be done in two different ways. First of all, we can think one thing without thinking something else, even if both are aspects of the same real thing. For example, we can think man without thinking white or six-feet tall. This is to abstract by way of simple apprehension. Second, we can think that one thing exists without another. This is quite a different mode of abstraction. It is abstraction by way of negative judgment. To abstract in the first way is simply to conceive of a nature without simultaneously being intellectually aware of anything else. It is to conceive a nature without being aware of those individuating and existential characteristics which are necessarily associated with that nature in real things. This is to think an object separately, but — and this is most important — it is not to think of it as though it were separate. As long as an object can be understood apart from something else it can be abstracted from it in this way. This is true even though the nature we know cannot exist apart from that from which it is abstracted. Not everything can be abstracted from everything else in this fashion. Some things are unintelligible apart from others and cannot even be understood apart from them, let alone exist apart from them. These, of course, cannot be abstracted from the others, not even in an abstraction by way of simple apprehension. Thus *animal* can be abstracted from *man* and *brute* because it presupposes neither although it is open to both. But *man* cannot be abstracted from *animal*, for *man* presupposes *animal* and cannot be understood apart from it.

An abstraction by way of negative judgment is a commitment to the existential separation of one thing from another. This is legitimate only when that which is thought to exist apart from something else does so exist. It is legitimate to abstract *man* from *stone* in this way, for the nature of stone and the nature of man never exist together in the same thing. However, it is illegitimate to think that something exists apart from something else when it does not, even though it can be understood apart from the other. Thus, even though *man* can be understood apart from *tall*, I cannot judge *This man is not tall* so long as this man actually happens to be tall. To recapitulate: one thing can be abstracted from another by way of simple apprehension as long as it is objectively (i.e., intelligibly) independent of the other, regardless of the existential relationship between the two; and one thing can be abstracted from another by way of negative judgment only as long as it is both objectively and existentially independent of the other.

The solution to our problem is clear. The object of simple apprehension is abstracted *from* individuating characteristics, and it exists only *with* individuating characteristics. However, it *can be understood* apart from them even though it cannot exist apart from them. The abstraction proper to the universal object is the first kind of abstraction, that is, abstraction by way of simple apprehension. However, abstraction by way of negative judgment would falsify the situation. The nature can be known apart from its individuating characteristics since it does not depend upon them for its definition. Yet it cannot be thought to be apart from them, for it does depend upon them for its existence.

The object of simple apprehension is never determinately the whole of any real thing. Yet a partial view of reality is not necessarily false. The sense of sight is not betrayed because it knows only the white of a white thing which is also sweet and moist. Neither is the intellect betrayed when it knows only an intelligible aspect of a thing which involves many additional characteristics.

III. The Division of the Universal Into the Predicables and the Categories

A concept can be classified from the point of view of its extension and from the point of view of its comprehension. From the point of view of its extension or universality the universal is divided into five types, because there are five different ways in which it can be said of its possible inferiors. In other words, there are five different ways in which a superior can be related to its inferiors. This does not mean that one concept can be related to the same inferior in five different ways. A given concept is related to an inferior in only one way, but another concept may be related to this inferior or to some other inferior in a second way, and so on until we have discovered five different ways in which a superior can be related to its inferior. Since these are the different ways in which a universal can be said of (i.e., is predicable of) many, these are called the five predicables. From the point of view of its comprehension the universal is divided into ten types. Obviously there are more than ten different concepts from the point of view of intelligible content. Yet if we consider the most general classification of things, that is, the division of being itself into its most universal types, we find that there are ultimately ten kinds of being. Every nature is basically one or other of these types. They are spoken of as the ten categories or the ten predicaments. The ten categories divide a concept directly or essentially, because they divide it according to differences in its essential content. The five predicables divide a concept indirectly or denominatively, because they divide it according to the different relationships which it takes on in the mind and which it bears in reference to its inferiors. To illustrate, let us anticipate for a moment. We shall see that man is ultimately that kind of nature which is substance, and that man bears the relationship of species to its inferior Peter. As to category man is a substance; and as predicable, it is a species (in reference to Peter). Man is essentially substance, since this is the ultimate classification of its actual intelligible content. Man is denominatively species (in reference to Peter), because the relationship it has (in reference to Peter) is precisely that of species.

We shall return to the division of the universal into the five predicables and the ten categories in the following two chapters.

Exercise IV

1. Define: universal, individual, universality, abstraction, abstraction by way of simple apprehension, abstraction by way of negative judgment, objec-

tive independence, existential independence, abstract concept, concrete concept, category, predicable.

2. What is the problem of the universal as it is manifested in the affirmative proposition with a singular subject?

3. Distinguish between the nature taken absolutely and the nature in its existential status.

4. Distinguish between the universal nature and the universality of the universal nature.

5. Solve the problem of the universal as presented in answer to No. 2 through the use of the distinctions given in the answers to No. 3 and No. 4.

6. Distinguish between abstraction by way of simple apprehension and abstraction by way of negative judgment. Under what circumstances is each abstraction legitimate? Under what circumstances would each falsify things?

7. What precisely is the difference between thinking something (to be) separate and thinking it separately? Which of these ways of thinking applies to the object of simple apprehension?

8. What is the difference between objective independence and existential independence? Which one implies the other?

9. What is the difference between a concrete concept and an abstract concept? Which of the following are concrete and which are abstract?
    9.1. Clever
    9.2. Dexterity
    9.3. Speed
    9.4. Measles
    9.5. Logic
    9.6. Logician
    9.7. Logical
    9.8. Honest
    9.9. Human nature
    9.10. Courage
    9.11. Rational
    9.12. Intellect
    9.13. Horrified
    9.14. Growth
    9.15. Growing
    9.16. Universal
    9.17. Relation
    9.18. Maple Tree
    9.19. Equal
    9.20. Action

    For each of these supply the correlative abstract or concrete expression. For example, *happiness* is abstract, and *happy* is its concrete correlative; *qualified* is concrete, and *qualification* is its abstract correlative.

10. From what point of view is the universal divided into the ten categories? From what point of view is it divided into the five predicables?

11. What is the difference between the division of the universal essentially taken and the division of the universal denominatively taken?

CHAPTER V

The Predicables

I. The General Notion of the Predicables

As noted previously, there is a difference between the nature or essence known in an act of simple apprehension and the relation of universality given to this nature by the mind. Furthermore, we have seen that it is in virtue of this universality that a nature can be said of its inferiors. An analysis of the content of the natures we know, that is, of what those natures are in themselves, yields the ten categories, i.e., the ultimate types of universal natures from the point of view of their essential content. An analysis of the universality of those natures, on the other hand, yields the five predicables, i.e., the five different ways in which a universal may be said of an inferior. For example, we shall classify *man* in *John is a man* categorically as *substance* and predicably as *species*. *Man* is in the category of *substance* because it is ultimately an instance of substantial being in terms of its meaningful content, even apart from its logical relation to John. *Man* — in this proposition — is related to John by way of the predicable *species*. *Man* is classified in itself or essentially as *substance*; it is classified in its reference to John or denominatively as *species*.

II. The Division of Universal Natures Into Five Predicables

By studying some examples we can see that some universals can be said of inferiors differently than others. *Animal*, *able to talk*, *my brother*, and *running* can all be said, i.e., are predicable, of John, but no two of these in quite the same way. John is essentially a man. Out of these four possible predicates only *animal* is within the comprehension of the concept *man*. The others are extrinsic to John's essence, and even they differ one from the other. *Able to talk* is predicable of John as something which belongs necessarily to all men, even though it is not within the comprehension of *man*; neither *my brother* nor *running* necessarily belongs to all men. Still *my brother* is not related to John in the same way as *running* is. John may be running now and not running later on, but once he is my brother he cannot cease to be my brother. These examples show clearly that universals differ considerably one from another as far as their relation to inferiors is concerned.

What then are these differences? What are the different predicables? First of all, a universal may be related to its inferior as something comprehensively intrinsic to its essence or as something extrinsic to its essence. If it is said essentially of its inferior it may express that essence in whole or in part. In the former case, we have the predicable known as *species*. This is the way *man* is said of John, for anything added to human nature in John is nonessential. If the latter is the case, the universal nature may be either the common and determinable or the proper and determining part of the essence. The universal which is predicable of its inferior as the common and determinable part of its essence is called its *genus*. This is the way *animal* is said of John, for *animal* is common to men and brutes and in itself can be more precisely determined as *rational* or *nonrational*. The universal which is predicable of its inferior as the proper and determining part of its essence is called its *difference*. This is the way *rational* is said of John, for *rational* is proper to (i.e., limited to) men and is that which determines *animal* in men by contracting the genus to the species.

If a universal is essentially related to its inferior it is, of course, necessarily related to it, for no subject can lack that which belongs intrinsically to its essence. Thus species, genus, and difference are all essential and necessary predicates. Whenever a universal is extrinsically related to a subject, it is a nonessential predicate. However, this is not to say that it is not necessary to that subject. A universal which does not belong to the essence of its inferior may be necessary to that inferior because it is an infallible consequent of the essence of that inferior. This type of universal is called a *property*. This is the way *able to talk* is said of John; although *able to talk* is not one of the comprehensive notes of *man*, it is an attribute necessarily present in any man. Of course not every nonessential predicate is necessary. Any universal which is predicated of a subject neither as one of its essential notes nor as a property necessarily connected with that essence is neither essential nor necessary. This type of universal is called an *accident*. *Running* is predicable of John as an accident, for it is neither intrinsic to his essence nor necessarily consequent upon it. John may be running, but not all men are running, and John need not be running and may, in fact, in a moment not be running.

THE PREDICABLES   43

These, then, are the five predicables, the five ways in which a universal can be said of its inferiors. The division of universal natures into five predicables can be schematically presented in the following way:

A universal may be said of its inferior either —

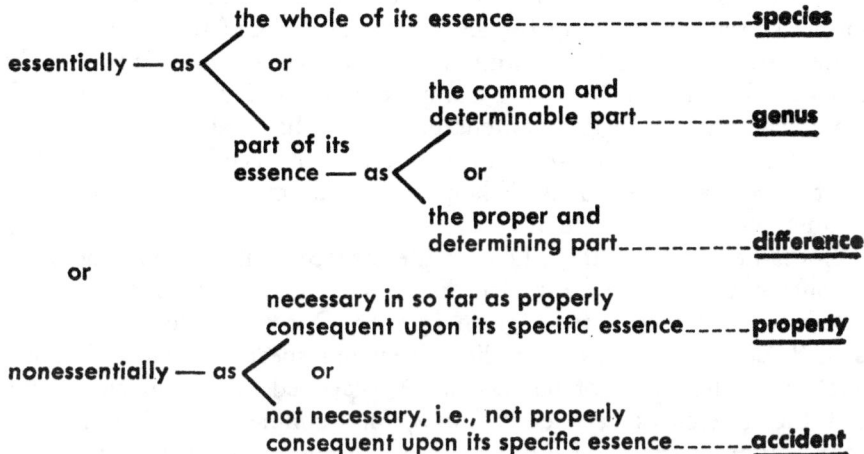

III. The First Predicable — Species

Species is defined as a universal nature predicable of many things which differ only nonessentially and numerically in answer to the question, "What is its essence?" Since the species of a subject is the whole of the essence of that subject, anything belonging to that subject over and above its species belongs to it in the manner of a nonessential or individuating characteristic. Hence, the only differences between the subjects of a given species will be nonessential. These subjects can differ only in so far as one is not the same subject as the other, although both are identically the same kind of subject. The species *man* can be said of Mary, Peter, and John, and of any other man. Essentially these subjects are all men, and they differ only in so far as one is not the same man the other is. In other words, as far as essence is concerned they differ only numerically.

Predicables are defined in terms of their immediate inferiors. The immediate inferior of the species is, of course, the individual. The only subjects of which any species is predicable are distinguishable only as different individuals within the same species. An apparent contradiction of this point arises in the case of those subjects of a species which represent subclasses under the species and hence can themselves be said of many. For example, *man* is predicable of *Eskimo* as its species, and *Eskimo* is in turn predicable of many singulars. Nevertheless, the proper subject of the species is the individual. Whatever is added to *man* to make *Eskimo* is a nonessential characteristic or an individuating note. *Man* is predicable of *Eskimo* and of any other subclass of men, just as it is of Mary, Peter, and John, as the whole of its specific essence. This is simply to say that what makes an Eskimo an Eskimo is extrinsic to the human nature which places him within the species *man*. However the concept *Eskimo*, precisely as a complex concept, includes within itself more than the intelligible content of *man*. What makes an Eskimo an Eskimo is accidental in so far as any Eskimo is specifically a man. But what makes an Eskimo an Eskimo is by no means accidental to an Eskimo precisely as Eskimo.

Since the species is the whole of the essence of its inferior, there can be only one species for any given inferior. It may be a difficult task to determine this species for a given inferior. Our experience of ourselves is firsthand and sufficient to indicate that our species is *man*. We know surely that the whole of our essence is explained in human nature and that the differences between Eskimos, for example, and other men are nonessential. But what of Fido? Is he specifically *brute*, or *dog*, or *spaniel*, or *cocker spaniel*, or what? What about a given individual bird? Is its species *brute*, or *bird*, or *sparrow*, or *English sparrow*, or what? Here our experience is not firsthand, nor perhaps is it sufficient for us to determine the species accurately in each case. As a matter of fact, it is not the business of the logician to determine the species of Fido or of some bird or even of himself. The logician discovers the *general* rule for determining the species of any given inferior. The logician points out that a predicate is the species of its subject only if it expresses the whole essence of that subject. But the task of determining the species in any given case is not the logician's; his task is properly that of determining what it is to be a species. This, he says, is to be that kind of universal nature which can be said of many things which differ only nonessentially and numerically. The point made here concerning the determination of species in a given case is true in reference to each of the predicables. The logician, as logician, does not classify concepts in different fields, but he does ascertain the instrument of classification by investigating the very nature of the predicables themselves.

IV. The Second Predicable — Genus

Genus is defined as that universal predicable of many things which differ in species, and, like species, in answer to the question, "What is its essence?" Unlike species, however, the genus does not include the whole essence but refers instead to the common and determinable part of the essence of its subject. Consequently, whatever is immediately added to any subject over and above its genus is an essential difference contracting the genus to a determinate species. Thus the immediate inferiors of the genus are species, e.g., *triangle* is a genus immediately predicable of *isosceles triangle, equilateral triangle,* and *scalene triangle.* Through these species, which themselves are immediately predicable of individuals, e.g., this or that triangle, the genus too is predicable of individuals. Hence, though it is defined in terms of inferiors differing in species, for these are its immediate inferiors, the genus is also predicable of subjects differing only in number. To give another example: the genus *animal* is immediately predicable of *man* and, through *man*, of Mary and Peter.

Unlike species, where there is only one to an inferior, there may, in a sense, be several genera to an inferior. The genus, as we have seen, is the common and determinable part of the essence of its subject. This common part may itself be wholly or partially expressed. The whole of the common part of the essence is called the proximate genus, because it is the genus closest to the species. The proximate genus lacks only that final essential difference which is necessary to determine the species fully. Thus *animal* is the proximate genus of *man* (and through *man* of Mary, John, and Peter). All that is lacking in the concept *animal* to make it the concept *man* is the final comprehensive note *rational*, which is what differentiates men essentially from all other animals. The concept *animal* includes the whole of the common part of the concept *man*, and thus *animal* is the genus closest to the species. But what of *organism, body,* and *substance* in reference to *man* or to Mary or Peter? Each of these is a genus, for each expresses a common and determinable portion of the essence of man. Yet they differ from *animal* in that in turn each expresses a more and more common and determinable part of the essence. *Animal* is common to men and brutes and must be determined by *rational* to become *man. Organism* is common to men, brutes, and plants and must be determined by *sentient* and then *rational* to become *man. Body* is still more common and further removed from *man.* Just as *animal* is called the proximate genus of *man* or Mary, so *organism* and *body* are called remote genera. *Substance* is the most remote or ultimate genus.

In the strict sense the species expresses the entire essence of its subject and is predicable properly only of individuals. In a less strict sense, we can say that species is formed whenever a genus is contracted by a difference. In this sense a species would be anything of which a genus is immediately predicable. This is to consider species from the point of view of its subjectibility rather than its predicability. Looking at species in this way, we can say that *animal*, in relation to the genus *organism*, is a species. If we do this, it is clear that what is a genus in one reference is a species in another reference and vice versa. Thus *body* is a species of *substance* and the genus of *organism*; *organism* is species of *body* and genus of *animal*; and *animal* is species of *organism* and genus of *man*. However, *substance* is only a genus and *man* only a species. Just as *substance* is called an ultimate or highest genus so *man* is, from this point of view, referred to as a lowest species. From this point of view it is also possible that what is a remote genus in one reference may be a proximate genus in another. Thus *organism* is remote genus of *man* or Mary, but proximate genus of *animal*. *Substance*, of course, is simultaneously proximate and ultimate genus of *body*.

V. The Third Predicable — Difference

Difference is defined as that universal which is said of many things differing either in number or in species in answer to the question, "What is the quality of its essence?" The difference is the proper and determining part of the essence, which, added to the genus, contracts the genus to a given species. When the difference is the final perfection of the essence, i.e., when it is the essential note added to the proximate genus to yield the species of its subject, we call it the specific difference. *Rational* is related to Mary as specific difference contracting the proximate genus *animal* to the species *man*. *Sentient*, which contracts *organism* to the proximate genus *animal*, is related to Mary as a generic difference. The same is true of *living* and *material* since they are differences respectively of *organism* and *body*.

The specific difference, like the species, is most properly predicable of the individual. However, it can also be said of the species by way of the individual. Thus *rational* is the specific difference of Mary and also of *man*. In the strict sense, of the four differences mentioned above, namely, *rational*, *sentient*, *living*, and *material*, only *rational* is a specific difference. However, just as some genera can in a loose sense be considered species (e.g., *animal* as species of *organism*), so the three generic differences of Mary and *man* can be considered specific differences, respectively of *animal*, *organism*, and *body*. Thus *sentient* is a generic difference for Mary and *man* but a specific difference for *animal*. *Living* is a generic difference for Mary, *man*, and *animal* but a specific difference for organism. *Material* is a generic difference for Mary, man, animal, and organism, but it is specific difference for *body*.

Thus far we have discussed the three essential predicables, namely, species, genus, and difference. We have seen that for any singular inferior there is but one species, one proximate genus, several remote genera, one ultimate genus, one specific difference, and several generic differences. Mary's species is *man;* proximate genus, *animal;* remote genera, *organism* and *body;* ultimate genus, *substance;* specific difference, *rational;* generic differences, *sentient, living,* and *material.* This is a complete list of the essential predicates for Mary or for any individual man. A similar list is difficult to determine for other subjects, with the exception of some mathematical entities, because our knowledge of their essences is less perfect. However, ultimate and remote genera and generic differences are fairly easily determined in all cases. Most of the serious difficulties arise in determining proximate genus, species, and specific difference. There is no doubt that for Fido *substance* is the ultimate genus, with *body, organism,* and *animal* as remote genera and *material, living,* and *sentient* as generic differences. It may be that *brute* is the proximate genus and that *dog* is the species. The specific difference probably cannot determinately be named.

VI. The Fourth Predicable — Property

Species, genus, and difference are essential predicates because they belong to the essence of their subject. But these do not exhaust the list of possible predicates. As a matter of fact, more often than not what we know about a given subject can be said of that subject only as something extrinsic to its essence. As we saw, there are nine predicates essentially related to Mary, each predicable either as species, genus, or difference. But there are probably hundreds of other predicates we could apply to Mary. We may know that she is studious, pious, pretty, John's cousin, able to talk, a mathematician, able to laugh, tired, hungry, nearsighted, artistic, biped, twenty-two years old, in Boston, etc. Each of these is a universal which can be said of Mary. But none can be said essentially, since none of them is within the comprehension of the concept *man.* Since each expresses something outside Mary's essence, each is predicable of Mary only nonessentially.

None of these attributes is intrinsic to Mary's essence, but some of them are so intimately tied up with her essence that they follow necessarily from it. These are attributes which belong to all men exclusively and always. Such nonessential but strictly necessary attributes are called properties of their subjects. Property can be defined, then, as that universal which is said of its inferiors nonessentially but with strict and exclusive necessity.

Neither Mary's ability to talk nor her ability to laugh is intrinsic to her essence. Neither makes her to be a man. But both follow infallibly from her essence, since they are necessarily caused by the principles proper to her essence, ultimately by her specific difference. She is a man not because she can talk and laugh, but she can talk and laugh precisely because she is a man. Every man can talk and laugh, men always can talk and laugh, and only men can talk and laugh. *Able to talk* and *able to laugh*, then, are universals related to Mary as properties. The intimate connection between the specific difference and the property should be carefully noted, along with the differences between them. Very frequently, as we have pointed out, it is difficult to determine the specific difference. In such cases, especially when we are attempting to define a concept, it is helpful to determine the properties of the subject. As long as we realize that they are extrinsic to the essence we can profitably use them to point to the essence, since they are connected necessarily with it. Both properties and specific differences belong always to all the members of a species and only to the members of the species. They differ in that the specific difference is the final note constitutive of the essence, while the property follows necessarily from the essence without in any way constituting it. A subject is a man because it is rational; it is able to laugh because it is a man.

Properties, strictly taken, like specific differences, are predicated of individuals, and through them of species. But, just as there are specific and generic differences, so too there are specific and generic properties. And just as there are differences generic in one reference and specific in another, so there are properties generic in one reference and specific in another. Just as *rational* is the specific difference of Mary because it differentiates her species from all other species, so *able to laugh* is a specific property of Mary because, although nonessential, it belongs always and only to every member of her species. And just as *material* is a generic difference for Mary because it differentiates the genus *body* from *spiritual substance*, so *corruptible* is a generic property for Mary because it belongs nonessentially always and only to every member of the genus *body*. Furthermore, just as *material* is a generic difference for Mary but the specific difference for *body*, so *corruptible* is a generic property for Mary but a specific property for *body*. In similar fashion, *having three angles equal to two right angles* is a generic property for *isosceles triangle*, since it is said nonessentially always and only of all the subjective parts of *triangle*, which is the proximate genus of *isosceles triangle*.

We might profitably suggest several other examples of property. *Able to fashion and use tools, able to take on the virtue of art,* and *possessed of natural logic* would each be a specific property for man. So similarly *having a boiling point of 100 degrees centigrade at sea level* is a specific property for water; *having three equal angles,* a specific property for equilateral triangle; and *interfertile only with members of that species,* a specific property for any given biological species. *Able to know sensible phenomena* would be a generic property for man; *capable of filling a place,* a generic property for water; *divisible into parts with a common boundary,* a generic property for equilateral triangle; and *able to support nonessential modifications,* a generic property for any biological species.

There is one word of caution in reference to the predicable property. To constitute a property it is not sufficient that an attribute belong *as a matter of fact* always and only to every instance of a given species. An attribute is a property only if it is so connected with the essence of a species that it is impossible for it not to be necessarily and exclusively an attribute of that species. A property is so related to its species that it would involve a contradiction to think of that species existing without the property. An attribute might, as a matter of fact, happen to belong always and only to every existing (past and present) member of a given species without being a property of that species. If the species could be thought of — without contradiction — possibly as existing without this attribute, it is something less than a property.

VII. The Fifth Predicable — Accident

Those universals which must be said of inferiors neither essentially nor with the necessity of properties are accidents. Accordingly, accident is defined as a universal said of many things nonessentially and contingently. Thus, *sitting, John's cousin,* and *biped* are all said of Mary as accidents, for none of these is intrinsic to her essence nor is any of these a necessary and exclusive property either of her species or genus.

In a sense accidents are the easiest of the predicables to determine. Yet even here we must distinguish the several kinds of accidents. Mary is sitting, but in a moment she may not be sitting. Mary is John's cousin and always will be, but John is not his own cousin. Mary is biped and so is every other man, and they always will be. Thus, though none of these has the necessity of a property, *John's cousin* is more necessary than *sitting,* and *biped* is more necessary than *John's cousin.* Since any man may at any moment either be sitting or not sitting, *sitting* may be classified as a wholly separable accident. *John's cousin*, on the other hand, is partly separable and partly inseparable. Inasmuch as some members of the species are John's cousins and some not, the attribute of being John's cousin is separable from the species; but since those who are John's cousins cannot cease to be so, this attribute is inseparable from the individual. *John's cousin*, then, may be classified as an accident separable from the species but not from individuals within the species. *Biped* is wholly inseparable, since no man can be without being biped. *Biped* falls short of being either a specific or generic property inasmuch as it is proper neither to the species (for birds are also biped) nor to the proximate genus (since dogs are not biped) nor to any remote genus (since it is not true even of all the members of the proximate genus). This subdivision of the predicable accident into wholly separable, partially separable, and wholly inseparable accidents is helpful in that it manifests the nature of accident in the variety of its instances. However, these differences are of little logical significance. The important point here is that all accidents, whatever their differences, are alike in that each is related to its inferior nonessentially and with something less than strict and exclusive necessity. This characteristic, common to all accidents, renders any accident of relatively little logical significance for purposes of definition or proof.

Examples of accidents, especially wholly separable accidents, are easily determined. When we say that Mary is pretty, or studious, or a mathematician, or in Boston, each time we predicate a separable accident of her. When we say she is feminine we predicate of her an accident inseparable from the individual but separable from the species. And when we say that she can breathe we predicate of her a wholly inseparable accident.

VIII. The Limitations of This Division

We have said that the five predicables are the five ways in which universals can be related to inferiors. Strictly speaking, the universals we refer to must be understood with certain restrictions. For a universal to be technically one of the predicables it must be a superior, i.e., an object which can be said of an infinite number of actual or possible inferiors in such a way that it has identically the same meaning for each. Not every predicate is a universal in this sense of the term. Some predicates are not related to their subjects as superior to inferior: for example, a predicate might be singular or it might be the definition of its subject. Other predicates are said of their subjects with something less than identically the same meaning in each case: we shall refer to these later as analogous predicates. Thus it is clear that the division of universal natures into five predicables is not so broad a division as to encompass all possible predicates. It applies in the strict sense only to a limited number of predicates. However, it is highly important division. First of all, the kind of predicate it does divide is of major concern to us. Second, those predicates which are, strictly speaking, outside its scope can be themselves classified by way of comparison with the predicables (as in the case of objects described as quasi-genera or quasi-properties, etc.).

IX. The Tree of Porphyry

The explicit listing of the predicables as species, genus, difference, property, and accident comes to us originally from the writings of Porphyry, an early commentator on the works of Aristotle. A schema representing the essential interrelations between concepts from the ultimate genus to the lowest species and even to the singular inferiors of this species has been reproduced through the centuries bearing the name of Porphyry. Because of its appearance it is called the Tree of Porphyry. It is helpful for students to use schemata of this type in order to see graphically the meaning and force of the predicables. The Porphyrian Tree, you should note, has to do only with the three essential predicables: species, genus, and difference. Use the schema in conjunction with the explanation of these three predicables as given above in sections III, IV, and V of this chapter.

THE TREE OF PORPHYRY

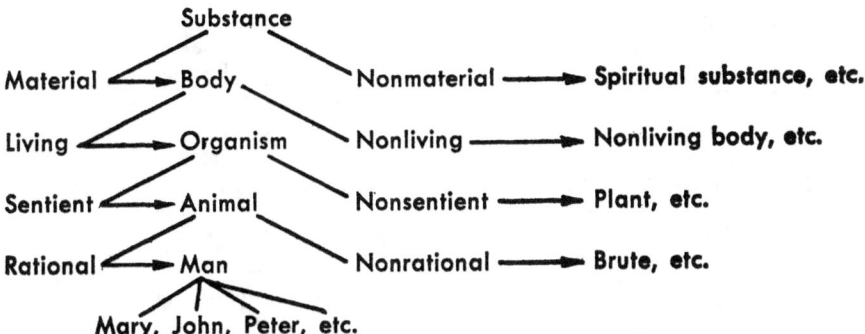

Exercise V

1. Define: species, genus, proximate genus, remote genus, ultimate genus, difference, specific difference, generic difference, property, specific property, generic property, accident, wholly separable accident, wholly inseparable accident, partly inseparable accident.
2. The predicables directly or essentially divide universality; indirectly or denominatively, the universal nature. Explain.
3. Explain why the predicables are divided into five.
4. What does it mean to say that species is said of many which differ *only in number*?
5. What part precisely does logic play in determining whether a predicate is species (or genus, or difference, etc.) of its subject?
6. Genus is defined as that which is said of many differing *in species*. Yet *animal* is genus for both Mary and Peter, and they differ *only in number*. Explain.
7. How do the several types of genera differ from one another although they remain alike in so far as each is a genus?
8. Why is difference defined as that which is said of many differing *in number or species*?
9. Both property and difference are said *always of all* of the members of a given species or genus and *only* of these. How do they differ?
10. How does accident in general differ from property, and how do accidents differ among themselves?
11. What is wrong with speaking of the predicables as the five ways in which a predicate can be said of its subject?
12. Practice reproducing the Tree of Porphyry, naming each entry in its reference to the entries above, below, and alongside of it.
13. Name the predicable relationship referring the objects on the right to those on the left in each case:
    13.1. Logic .... art
    13.2. Mary .... tall
    13.3. Body .... material
    13.4. Genus .... predicable
    13.5. "Horse" .... sign
    13.6. Fido .... boxer
    13.7. Fido .... four-legged
    13.8. Fido .... sentient
    13.9. Fido .... corruptible
    13.10. Fido .... hungry
    13.11. Fido .... male
    13.12. Fido .... son of Spotty
    13.13. Fido .... substance

THE PREDICABLES

13.14. Triangle .... three-sided
13.15. Triangle .... possessed of three angles equal to two right angles
13.16. Mary .... artistic
13.17. Mary .... rational
13.18. Maple tree .... plant
13.19. Intellect .... faculty
13.20. Piety .... virtue
13.21. Triangle .... plane figure
13.22. Candy .... sweet
13.23. Coal .... black
13.24. Organism .... living
13.25. This rock .... large
13.26. This rock .... body
13.27. This rock .... extended in space
13.28. This maple tree .... belongs to John
13.29. This maple tree .... can function vegetatively
13.30. Species .... predicable

14. Suggest as many predicates as you can think of for any given subject, and then classify each as to predicable relation.

CHAPTER VI

The Categories

I. The General Notion of the Categories

There is a distinction between the nature known in simple apprehension and its relation of universality to many inferiors. Our analysis of the ways in which universal natures can be related to their possible inferiors has revealed the five predicables. Now we shall consider the different ways in which universal natures can be classified from the point of view of their comprehensive content. This analysis will reveal the ten categories. The ten categories are the ultimate types of natures. This is to say that they are the most general classifications of being. They divide possible predicates on the basis of fundamental differences in intelligible content. They are also spoken of as the ten "predicaments," a name taken from the Latin word for predicate, just as the word "category" is taken from the Greek word for predicate. The difference between the predicables and the categories can be seen in terms of two different questions (and their answers) which can be asked about any universal concept. The answer to the question, "What kind of being is known in this concept?" will be given as one of the ten categories. The answer to the question, "How is this concept related to a given inferior?" will be given as one of the five predicables. To repeat an example given before in the same context: man considered in its intelligible content is classified in the category substance, because ultimately it is that kind of being which is apt to exist in itself and not in another; *man* considered in its logical relation to Mary or any other individual man is classified predicably as species, for it expresses the whole essence of Mary and of all individual men.

II. Preliminary Considerations

Before discussing the division of being into its ten categories, there are several preliminary points to consider. The division is not nearly so simple as it might seem at first glance, and these points must be clear before the division can be understood. Aristotle himself prefaces his work on the categories with several distinctions, which have come to be known as the antepredicaments, that is, "before the predicaments or categories."

First of all we must distinguish carefully between univocal, equivocal, and analogous names or terms. Terms are signs of concepts and as such signify meanings. A term is said to be used univocally if it is used several times to refer to identically the same meaning. To say "man" of Mary and John, to say "animal" of Mary and Fido, and to say "triangle" of an isosceles triangle and an equilateral triangle is each time to use a term univocally. Mary and John are equally man. Mary and Fido are equally animal. And the isosceles triangle and the equilateral triangle are equally triangle. However, the same term may be used several times to refer to utterly unrelated meanings. When this happens the term is said to be used equivocally. Thus I speak equivocally when I use the same term "sound" to refer both to the object of the sense of hearing and to a long passage of water joining two large bodies of water. It is purely by chance that we use the same name in these two instances. This is also the case when I call a certain instrument for writing and an enclosure for young children a "pen" or the external covering of a tree and a small sailing vessel a "bark." Sometimes one term may stand for different but related meanings. Such a term is said to be used analogously. An analogous term is always said in a primary sense of one of its subjects and in a secondary sense of the others. To understand what the analogous term means in reference to the subjects of which it is said secondarily we must understand what this term means when predicated of its primary subject. Thus the term "proud" can be said primarily of certain men and secondarily of peacocks. It does not mean the same thing in each case, but there is a relationship between the two cases so that we understand the second use by referring to the first use. Likewise the term "healthy" is primarily said of physical organisms and secondarily of the weather, "talking" is primarily said of men and secondarily of parrots, "good" is primarily said of God and secondarily of creatures. As a matter of fact there are significant differences in analogous terms. However, we can leave the division of analogy into its types for consideration elsewhere.

Our second preliminary consideration is the distinction between complex and incomplex objects of thought. Here it is sufficient simply to review what we have already seen. In one sense the object of simple apprehension differs from the object of judgment as the incomplex from the complex. In simple apprehension we merely grasp a nature or quiddity. In judgment we know not only quiddity but the act of existing, for we make an affirmation or denial about the way a nature *is*. In another sense one object of simple apprehension may differ from another as the incomplex from the complex. From the point of view of what is known, *man* and *stone* are both incomplex objects of simple apprehension, while *healthy man* and *pebble* are both complex. *Man* involves only human nature and *stone* only stoneness. But *healthy man* involves both human nature and health, while *pebble* involves both stoneness and smallness.

Our third preliminary consideration is twofold. It has to do, first, with the distinction between the universal and the singular. The universal (e.g., *man* or *dog*) is an object which can be said of many. The singular (e.g., *Mary* or *Fido*) is incommunicable and can in no way be said of many. Second, we must distinguish between that kind of a nature which is such that it can exist in itself and not in another and that kind of nature which is such that it can exist only in another as in a subject. This is the distinction between substance (e.g., *man* or *dog*) and accident (e.g., *tall, white, hungry*). Height is always the height of something, and color is always the color of something. Height and color exist only in so far as they inhere in a subject. They are accidents. One accident may inhere in another. But there can be no accident whatever unless something exists in itself and does not inhere in a subject. An infinite series of accidents is an absurdity. Human nature is the kind of being which exists in itself and is ultimately the subject of accidents. Human nature is a substance. Combining these distinctions we can differentiate between singular substances, universal substances, singular accidents, and universal accidents. Singular substances (called primary substances) neither exist in a subject nor are said of a subject, e.g., *Mary, Fido, this maple tree*. Universal substances (called secondary substances) do not exist in subjects but are said of subjects, e.g., *man, dog, maple tree*. Singular accidents exist in a subject but are not said of a subject, e.g., *Mary's kindness, Fido's hunger, this maple tree's greenness*. Universal accidents, finally, both exist in subjects and are said of subjects, e.g., *kind, hungry, green*.

A final preliminary distinction is that between real being and being of the reason. Real being, in contrast to a being of the reason, either does (actual real being) or does not but can (possible real being) exist outside the mind. On the contrary, a being of the reason is a being which can exist only as known. It is contradictory to think of a being of the reason existing anywhere except in the intellect as an object known. Men, dogs, and maple trees are examples of actual real being. A better world and a thirty-five-year-old President are examples of possible real being. All logical relations such as genus or species, fictitious beings like two-headed dragons, negations and privations such as blindness or deafness, and some mathematical entities such as the square root of a minus number are examples of beings of the reason. Men, dogs, and maple trees may exist in the mind as known, but they also exist outside the intellect quite independently of our act of knowing them. A better world and a thirty-five-year-old President actually exist only in the mind as known, but there is nothing in their natures as known which would make existence for them outside the intellect a logical impossibility. However, genus, blindness, and the square root of minus three exist only in the intellect as known and are such that it would be contradictory for them to exist in any other fashion. Only the latter are beings of the reason; the others are real beings, actual or possible.

III. What Kind of Being Can Be Categorized

When we speak of being as divisible into its ten ultimate types, which are the categories, we must be careful to note that only certain kinds of being can be categorized. Not all being fits into one or other of the categories. Before investigating the division of being into the categories, then, we should consider these restrictions, for no division makes sense unless we know precisely what it is we are dividing. These restrictions can be listed as six in number:

1. Only real being is categorized. Only real being, that is, being which does or can exist outside the intellect can be categorized. Thus beings of the reason such as logical relations (e.g., genus), fictional beings (e.g., chimera), certain mathematical entities (e.g., square root of minus three), privations (e.g., blindness) and all other beings which exist only as known are not categorized.

2. Only incomplex natures are categorized. This means, first of all, that only objects of simple apprehension can be categorized. The objects of judgment, since they are complex, cannot. Second, it means that only the incomplex object of simple apprehension can be categorized. Objects of simple apprehension which are complex in themselves (e.g., healthy man or pebble) are composites of two or more natures; each of these in its own right fits into a given category, but the complex object formed by their union does not. To anticipate for a moment, man is categorized as substance and health as quality. Healthy man, therefore, cannot itself be classified in any one category. The same is true of any artificial nature (e.g., table or pencil), for all artificial natures are complex, since they are composed of a material principle (one nature) plus an added artificial form (a second nature).

3. Only finite being is categorized. The categories are particular modes of being, each representing a determinate limitation or contraction of being. Infinite being defies limitation and cannot be categorized.

4. Only univocal terms are used in classifying natures according to category. All equivocal terms and some analogous terms stand for a plurality of entirely different natures. These natures may fit into a variety of categories. This is the case when "bark" is used equivocally in relation to trees and dogs. The bark of a tree is categorically substance, whereas the bark of a dog is categorically action. This is also the case when "foot" is used analogously in reference to a part of the human body and to a rank in a class. The foot of a body is categorically substance, whereas the foot of a class is categorically relation. Some analogous terms do not stand for a sheer plurality of entirely different natures. For example, "good" can be said of a man and also of his complexion. In each case "good" stands for something different, but for things which are nevertheless proportionally alike. It is not the same thing for a man to be good and for his complexion to be good. But each is good precisely in so far as each is what it should be. Analogous terms such as "good" stand for a perfection which is called transcendental because it "spills over" from one category to another. Since this is the case it cannot be said to belong to any one category.

5. Only complete natures are categorized. Those things which are called beings only in so far as they are parts of a whole which is properly the being in question are not beings; hence they are not categorized. In so far as they are parts of a whole, however, they can be reduced to the category of the whole. Because man is in the category of substance we can classify his parts as substance. Thus his hair, his teeth, his blood, etc., are classed as substance. This is also true of his body and soul considered as parts of his nature. The comprehensive notes of any concept, sometimes called metaphysical parts, are categorized like that concept. When expressed as genera they belong directly to the category, and when expressed as differences they belong indirectly to the category. Thus, *animal* and *organism*, like *man*, are directly in the category *substance*; while *rational*, *sentient*, and *living* are indirectly categorized as *substance*.

6. Only universal natures are categorized. In the strict sense the categories are classifications of possible predicates. As such they represent in the strict sense a division only of the universal nature. Since the content of a universal nature is actually found in individuals, the categories can themselves be said even of singulars. Thus, though John is not said to be categorized as substance, John is nevertheless a substance. The same cannot be said in reference to the predicables. Though *man* is species, it is not true that *John* is a species.

IV. The Division of Being Into Ten Categories

The categories are the ultimate kinds of being. They represent the basic types of being which we meet in experience. As such, they refer to the different kinds of intelligible content which is offered to our minds by the natures we know in simple apprehension. At the same time, the categories are the ultimate genera of being, that is, the broadest possible univocal predicates which can be said of subjects. The property of being a predicate is a logical property which finds its basis in the way its subject exists in the real order. This means that the modes of predication follow upon the modes of being and can be taken as reflections of them. Thus we can discover what the different categories are by considering the differences we find in predication. Let us see what an analysis of this kind shows. If the predicate expresses what the subject is, it is (1) substance; if it expresses a modification of the subject, it is an accident. If it is an accident, it will be taken either from something intrinsic to the subject or from something extrinsic to the subject. If it is taken from something intrinsic to the subject, it can be taken either from that which is absolutely intrinsic or from that which is only relatively intrinsic. An accident taken from something absolutely intrinsic will be taken either from the matter or from the form of the subject. That accident taken from the matter is (2) quantity; that taken from the form is (3) quality. The accident taken from what is relatively intrinsic to the subject is (4) relation. An accident taken from something extrinsic to the subject can be taken from something considered either under the aspect of a cause or under the aspect of a measure or under the aspect of neither. That accident belonging to a subject as cause of an effect in something extrinsic is (5) action; while the accident belonging to a subject as suffering an effect brought about by something extrinsic is (6) passion. Subjects are extrinsically measured by time or place; thus, the accident belonging to a subject because of an extrinsic measure will be taken either from time or from place. The accident belonging to a subject because of the time in which it is situated is (7) when. The accident arising in a subject simply because of the place in which it is found is (8) where, and because of the precise order or disposition of its parts in this place, (9) position. Finally, the accident belonging to a subject because of something wholly extrinsic to it which is neither a cause nor a measure, e.g., its clothing, is (10) having possession.

This division into ten categories can be seen in the following schema:

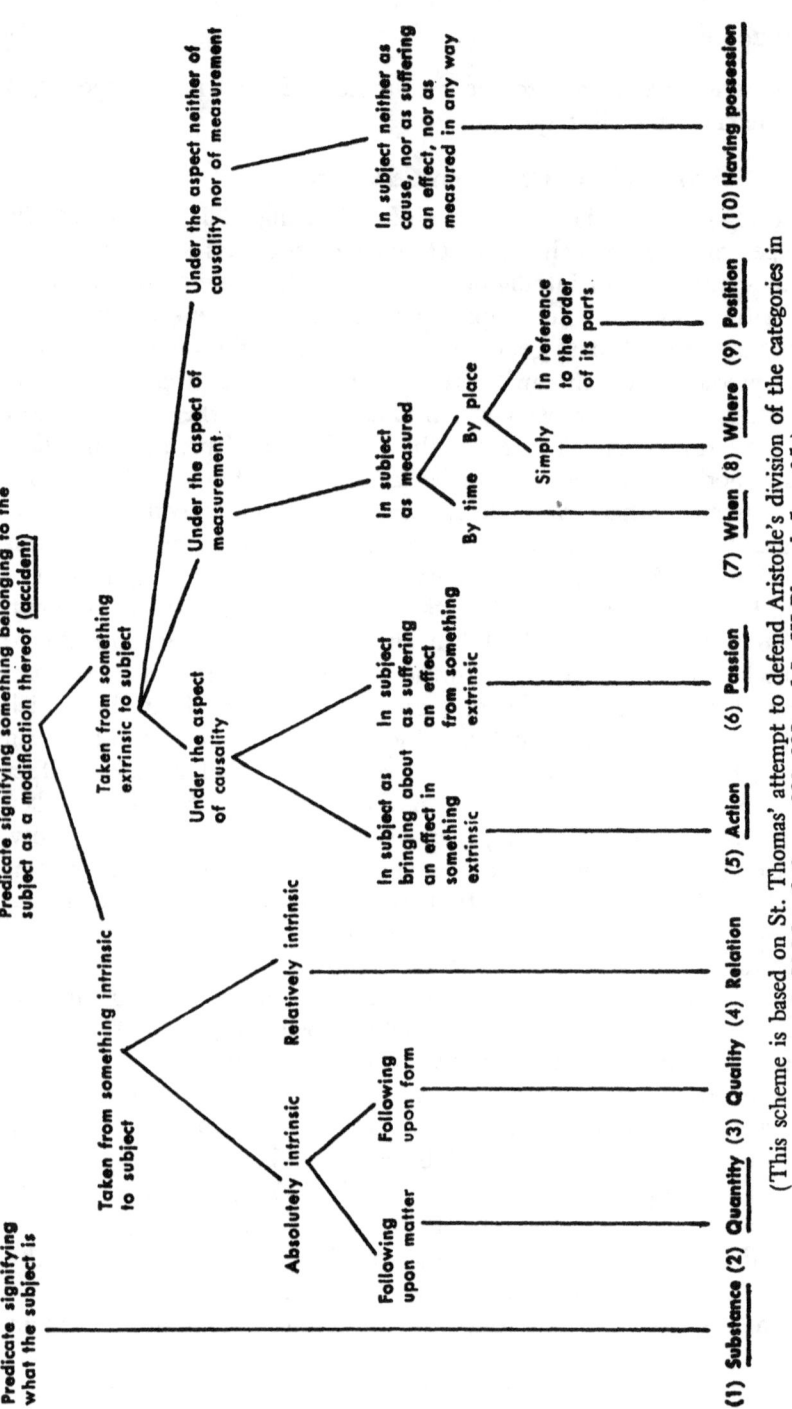

Since a predicate says something about its subject, we can think of the categories as answers to ten ultimately different questions that can be asked about a subject. These questions can be formulated in this fashion:
1. What is the subject? (*Substance*)
2. How much (or how many) of this subject is there? (*Quantity*)
3. What kind of a subject is this? (*Quality*)
4. How is this subject referred to other things? (*Relation*)
5. What is this subject doing to something else? (*Action*)
6. What is being done to this subject by something else? (*Passion*)
7. When is this subject doing something or being acted upon? (*When*)
8. Where is this subject? (*Where*)
9. What is the posture of this subject? (*Position*)
10. How is this subject dressed, adorned, or otherwise artificially equipped? (*Having possession*)

Before discussing each of the categories in detail, we should clarify one point concerning this division. The categories themselves are univocal. This means that all those things which are in the category of substance are equally substances, i.e., that *substance* means the same thing for each of them. The same is true of *quantity, quality, relation,* and the others. Yet these ten categories are themselves only analogously related to *being.* Because *being* does not mean precisely the same thing in reference to each of them, they are spoken of as particular modes of being rather than species of the genus *being.* "Genus" properly refers to a universal which can be divided into species through opposing differences which add something to the genus. Thus *animal* is contracted to *man* through the addition of *rational,* a note different from and determinative of *animal,* and to *brute* through the addition of *nonrational,* similarly a different and determinative note. But nothing can be added to *being* which is different from being save *nonbeing,* which is, of course, nothing. Hence there is no difference which could contract *being* as a genus to a species inferior to it. The categories differ from one another, but the very differences between them *are,* i.e., are *beings.* Hence *being* already includes the differences which distinguish its different modes and, accordingly, already includes the different modes themselves. *Being* can be considered a superior in reference to the ten categories, but not in the way in which we have seen that a predicable is a superior. A univocal superior, that is, a predicable, does not actually contain its inferiors; it includes them only potentially. Being, on the other hand, is an analogical superior or analogical whole. This means that *being* actually includes its inferiors and that they are subjective parts in only a loose sense. Being is not a genus; it is at best only a quasi-genus. We cannot here examine this extremely important point. However, we can note that *being*, although analogically superior to *substance* and all its inferiors, is not, like a univocal superior, related to these inferiors as a concept which has greater extension and less comprehension. *Being*, in fact, is both the most extended and, in a sense, the most comprehensive of concepts. The law of the inverse ratio between comprehension and extension applies strictly only to univocal concepts. Furthermore, we have noted that *being* is not a genus. Consequently, the categories are, as we have said above, the ultimate genera. Finally, even the differences between the various modes of *being* are themselves *being*. This makes it extremely difficult to define the categories in such fashion as clearly to show how they do differ from one another as distinct modes of *being*. This last observation leads us to the next section in this chapter where we shall attempt to define the categories and give examples of each.

V. The Ten Categories Considered Singly

1. Substance

Substance is defined as that nature which is apt to exist (i.e., calls for existence) in itself and not in another.

Examples of substance are *man, maple tree, horse, iron, water*. These are substances precisely because they are the types of natures which call for existence in themselves, not in another. This does not mean they exist by themselves, for each possesses the property of upholding accidents which inhere in it as nonessential modifications. No substance — save God, and God, as infinite being, cannot be categorized — exists without some accidents. These accidents are beings which call for existence in another as in a subject. Accordingly they are sometimes referred to as beings-of-a-being, while substance is being simply speaking.

2. Quantity

Quantity is defined as that accidental modification of a subject which causes it to have parts outside of parts.

Examples of quantity are *six feet tall, sixty-four pounds, two quarts, seventeen*. Quantity is the first accident arising in material substance, and it is that accident whereby the subject can be divided into constituent parts and measured by number. Quantity is divided into continuous quantity or extension and discrete quantity or number. Discrete quantity is derived from continuous quantity by the separation of one part of continuous quantity from another. Since it is the first accident arising in material substance, quantity is an accident through which other accidents inhere in material substance (e.g., color by way of surface).

3. Quality

Quality is defined as that accidental modification of a subject which causes it to be determined in itself in a certain fashion.

Examples of quality are best understood if we first point out four types or species of quality. These are (1) perfections of a substance disposing it well or badly in its being (entitative qualities) or in its operation (operational qualities) which are either firmly founded (habits) or are easily lost (dispositions); (2) abilities, i.e., potencies for operation, including their deficiencies but not the lack of them, i.e., their privation; (3) passive sensible characteristics which affect the sense powers; and (4) the qualitative termination of quantity known as figure or, if it includes aesthetic proportions, form. Examples of the first species include *healthy, vicious, mathematician, even seeing, understanding* and *loving,* as we shall see later. Examples of the second are *able to laugh, able to see, masculine, feeble-minded,* and *nearsighted.* Examples of the third are *red, sweet, pungent,* and *cold.* Examples of the last are *curved, triangular,* and *oblong.*

4. Relation

Relation is defined as that accidental modification of a subject which consists simply in its reference to another.

For a fuller understanding of relation we should distinguish between the relation itself (e.g., *paternity*), the subject of the relation (e.g., *John*), the term to which the subject is referred by the relation (e.g., *John's son*), and the foundation of the relation (e.g., *John's generative act in begetting his son*). With these distinctions in mind we can suggest the following examples of relation: *father, near, taller, teacher, equal.*

5. Action

Action is defined as that accidental modification of a subject which belongs to it precisely as the principle of a transitive operation.

Transitive operations differ from immanent operations in so far as they begin within an agent and pass from it to an extrinsic patient. Immanent operations remain within the agent for its own perfection. Pushing is an example of a transitive operation and seeing is an example of an immanent operation. Immanent operations qualitatively modify their subjects, whereas transitive operations modify their subjects in the category of action. Examples of action include *striking, pulling, lighting, injuring, helping,* and *tying.*

6. Passion

Passion is defined as that accidental modification of a subject which belongs to it precisely as the term of a transitive operation.

Examples of passion include *being struck, being pulled, being hit, being injured, being helped,* and *being tied.*

7. When

When is defined as that accidental modification of a subject which belongs to it precisely because it is measured by time.

Examples of when include *today, in 1492, last month,* and *in the Middle Ages.*

8. Where

Where is defined as that accidental modification of a subject which belongs to it precisely because it is measured by place.

Examples of where include *in Massachusetts, at home, in the kitchen, down South,* and *on the ball field.*

9. Position

Position is defined as that accidental modification of a subject which belongs to it precisely because of the disposition or arrangement of its physical parts in a given place.

Examples of position include *stooped over, standing erect, bent, sitting down,* and *at ease.*

10. Having possession

Having possession is defined as that accidental modification of a subject which belongs to it precisely because it is clothed, armed, or otherwise possessed of artificial trappings and equipment.

Having possession is perhaps the least easy of the categories to understand. Strictly speaking, only man is subject to this accident. Other subjects have it only under the care of men. Since man is a rational being, it is neither fitting nor necessary that he be substantially equipped with all that he needs for survival and for the happy life, because he can devise by way of art the clothes, ornaments, and arms he needs to lead the full life. When he actually possesses these exterior trappings, man is accidentally perfected by way of having possession. Examples of having possession are *being dressed, wearing earrings, armed to the teeth,* and *shod.*

Do not confuse the accident of having possession with a man's clothing or where with place or when with time. Having possession is an accident belonging to a subject because of his clothing; where, because of the place he is in; and when, because of the time. But in no instance is the accident to be confused with the thing from which it is taken.

VI. The Categories and the Predicables

It is important to keep in mind the distinction between the categories and the predicables. The categories divide natures absolutely or in terms of their comprehensive content. The predicables divide natures on the basis of the universality referring them to their inferiors. Any univocally universal concept, as long as it fits the requirements for categorical being, can be classed both as a category and as a predicable. For example, take the proposition: *Mary is sitting down.* Here the predicate *sitting down* is in the category of position and, as a predicable, is related to Mary as a wholly separable accident. Or take: *Fido is an organism.* Here the predicate *organism* is a substance predicably related to Fido as a remote genus. In some cases predicates defy classification as categories because they do not meet all the requirements for categorical being. These predicates may, however, allow for classification as predicables. As an example, take the proposition, *John is deaf.* The predicate *deaf* is a privation and consequently cannot be categorized. Yet *deaf* is predicably related to John as a wholly separable accident. Similarly in the proposition: *Genus is a logical relation*, the predicate *logical relation* is not a real being. Hence it cannot be put into any category. Yet it is predicably related to its subject as its remote genus. In addition, we have seen that complex natures are composed of a plurality of different natures. As such, they cannot be categorized. However, a predicate representing a complex nature may be used primarily to call attention to only one of the perfections it represents. The predicate of such a proposition can be classified according to the category of that perfection. Take *Aristotle was a philosopher* as an example. Since *philosopher* here formally, i.e., primarily, designates Aristotle as one *possessed of philosophy* and not as man, *philosopher* is put in the category of quality. As a predicable, it is related to Aristotle as a wholly separable accident.

It is particularly important, in view of the terminological identity, to keep distinct the notions of accident as a category and as a predicable. A nature is categorically an accident if it can exist only in another and not in itself. A nature is predicably an accident when it is related nonessentially and contingently to its subject. Categorical accidents need not be predicably accidents. In *Mary is able to talk, Running is an action,* and *This figure is an isosceles triangle,* the predicates are all categorically accidents but are predicably related to their subjects as specific property, ultimate genus, and species respectively.

It may be worthwhile at this time to say something about the use of the term *essence*. As a matter of fact, every nature, substantial or accidental, is an essence. It may seem that we have sometimes restricted the use of "essence" to substantial natures, because the examples we have used so far for the most part have involved substances, especially primary substances, as subjects. *Able to talk* is nonessential to Mary, considered as a primary substance, precisely because it is outside of her essence *as a substance.* This does not mean that *able to talk* is not an essence in itself. It is, of course, that kind of accidental essence classified as quality. Similarly, *virtuous* is said nonessentially of John, but *virtue* is itself an essence and is predicated essentially of John's temperance. With this in mind we see that substance is not the only category which can be divided by essential differences into branches to form a tree-like logical schema. Porphyrian Trees are theoretically possible for all ten categories, beginning in each instance with the ultimate genus in question and descending by way of essential differences to the singular instances in the category in question.

VII. The Modes of Opposition

The categories are studied in logic in order that we may see how different concepts are related one to another. Hence it is worthwhile to note how some concepts are opposed to one another, for opposition is a kind of relation. Accordingly the modes of opposition are traditionally considered immediately after the categories. Just as those things considered as preliminaries to the study of the categories are known as antepredicamental, so the modes of opposition are spoken of as postpredicamental.

Two concepts are opposed to one another when they are so related that the two cannot be realized in the same way at the same time in the same subject. There are four different modes of opposition. In two of these, one concept is positive and its opposite negative. In the other two, both concepts are positive. The two negative modes of opposition are those of *contradiction* and *privation;* the two positive modes, those of *contrariety* and *relation.*

The strictest form of opposition is that of contradiction. Contradiction arises between two concepts when one is the pure and simple denial of the other. This is the kind of opposition between an affirmation and its denial, and it can be reduced to the opposition between being and nonbeing. Contradictories are so purely opposed that one or the other is predicable of any subject whatsoever: if one of the contradictories cannot be predicated of a subject, the other must. Thus, if a subject is not white (whether or not it ever was or could be white), it must be nonwhite. *White* and *nonwhite* are contradictories. Thus whenever *white* cannot be said of a subject (e.g., coal or grape juice), *nonwhite* must (for certainly if neither coal nor grape juice is white, both are nonwhite). Other examples of contradictories are *logical* and *nonlogical*, *virtuous* and *nonvirtuous*, *higher* and *nonhigher*. The contradictory of any term is easily formed simply by prefixing "non" to the term. The resultant term expresses the pure denial of the first and is predicable of, and only of, every subject of which the first is not predicable. Contradictories can, of course, be expressed without use of the prefix "non," just so long as the negative term expresses the pure denial of its opposite. This is the case with *material* and *immaterial*, for *immaterial* has the force of *nonmaterial*.

The second type of negative opposition is that of privation. This mode of opposition exists between two concepts when one expresses a certain positive perfection or formality and the other expresses the lack of this perfection in a subject which could possess it. The difference between contradiction and privation is immediately evident. Contradictories are predicable of all subjects in the sense that the one must be said of any subject of which the other cannot. Concepts related by privation, however, are predicable only of a certain type of subject. The privative concept is predicable only of the subject which could possess its opposite. *Sight* and *blindness, sincerity* and *insincerity, legibility* and *illegibility* are examples of concepts related by privation. Blindness is the lack of sight in one who could see. Insincerity the lack of sincerity in one who could be sincere. And illegibility is the lack of legibility in something which could be legible. Thus *blind, insincere,* and *illegible* are said of subjects when *seeing, sincere,* and *legible* cannot be said of them. But they are not said of all subjects which are non-seeing, non-sincere, and non-legible. They are said only of those subjects of which these *could* be said because the subject could have sight, sincerity, and legibility. Notice that wherever a concept has a privative its contradiction can be said of every subject of which its privative can be said, but not vice versa. For example, *nonsincere* can be predicated of all the subjects of which *insincere* can be predicated. But *insincere* cannot be predicated of all the subjects of *nonsincere*. The reason for this is that the privative concept says more than the contradiction. Like a contradictory concept, a privative-concept expresses the denial of its opposite. In addition, and unlike the contradictory concept, it expresses the absence of its opposite in a subject in which that opposite could be found.

The first type of positive opposition is that of contrariety. Contraries are positive extremes in a given genus. Both of the concepts related by contrariety are positive, each expressing an extreme in reference to the other. Thus *black* and *white* are *contraries*. They are opposed as opposite extremes within the genus *color*. Contraries like concepts related by privation are predicable only of a certain type of subject. Colors belong to substances precisely in so far as these substances are corporeal and have a surface. Hence, both *black* and *white* can be said of, and only of, bodies. Some contraries, such as *black* and *white*, are mediate, since they admit of intermediary concepts (e.g., *gray*). Other contraries, such as *odd* and *even*, which are said of whole numbers, admit of no intermediaries and are called immediate contraries. In the case of immediate contraries, as long as we stay with the appropriate subject, what one contrary cannot be said of, the other can be said of. This is not so with mediate contraries. Thus, if a whole number is not odd it has to be even; but a piece of paper may be neither black nor white, though it could be either. Notice once again that the contradiction of a concept can be said of any subject of which the contrary of this concept can be said, but not vice versa. The contrary too says more than the contradiction. It expresses not only the denial of its opposite but it affirms the positive extreme of this opposite. It is sometimes difficult in practice to distinguish a privative from a contrary. The difference, of course, lies in the fact that a privative is negative and a contrary positive. However, some terms sometimes express a positive notion and at other times a negative notion. *Bad* can express either the privation of *good* or a positive extreme, depending on its use. Other pairs of concepts like *merciful* and *merciless*, *healthy* and *sick*, and *kind* and *unkind* present a similar difficulty.

The second type of positive opposition is the opposition between correlative notions which imply one another in different subjects. This is the opposition of relation. Correlative notions are defined in terms of the relation between them; hence one always enters into the understanding of the other and implies the other. They are opposites precisely because one implies the other in a subject distinct from its own subject. What is higher implies what is lower, for to be higher is always to be higher than what is lower. But what is higher cannot in this same reference be what is lower. Examples of relative concepts include *parent* and *child*, *husband* and *wife*, *taller* and *shorter*, *teacher* and *student*, *offensive* and *defensive*, *half* and *double*, *upper* and *lower*, and *front* and *back*. Here again we note that where one of two correlatives can be said of a given subject the contradiction of the other can also be said, but not vice versa. The relative expression also says more than the contradiction. In addition to the denial of its opposite in its own subject it implies the existence of its opposite in another subject.

Exercise VI

1. Define: predicament, univocal term, equivocal term, analogous term, substance, accident, real being, being-of-the-reason, quantity, quality, relation (as a category), action, passion, when, where, position, having possession, primary substance, secondary substance, continuous quantity, discrete quantity, habits, dispositions, transitive operation, immanent operation, opposition, contradiction, privation, contrariety, relation (as a mode of opposition).
2. Precisely how do the categories differ from the predicables?
3. Classify the following terms as to univocity, equivocity, and analogy:
    3.1. ...... "Fast" as signifying a runner and a color.
    3.2. ...... "Slip" as signifying a trivial error and a case for a pillow.
    3.3. ...... "Virtue" as signifying temperance and fortitude.
    3.4. ...... "Right" as signifying the opposite of left and correct.
    3.5. ...... "Quantity" as signifying continuous and discrete quantity.
    3.6. ...... "Creator" as signifying God and an artist.
    3.7. ...... "Head" as signifying part of the human body and the leader of a group.
    3.8. ...... "Water" as signifying a lake and an ocean.
    3.9. ...... "Cabbage" as signifying a vegetable and money.
    3.10. ...... "Lamb" as signifying a certain type of brute and Christ.
    3.11. ...... "Science" as signifying arithmetic and geometry.
    3.12. ...... "Medical" as signifying a physician and his surgical instruments.
    3.13. ...... "Beautiful" as signifying God and a sunset.
    3.14. ...... "Lark" as signifying a type of bird and a frolic.
    3.15. ...... "Bright" as signifying a light and a student.
4. Natures as known exist in the intellect. Why is "nature known" not synonymous with "being-of-the-reason"?
5. Keeping in mind the restrictions which make a being capable of being categorized, give several examples of predicates which do not fit into any category.
6. Explain the division of being into the ten categories.
7. *Substance* and the other categories are true genera, while *being* is at best a quasi-genus. Explain.
8. Check the following literary passages for as many examples of the ten categories as can be found in them. Do the same thing with passages

taken from textbooks and reference books for other courses, especially those of a scientific nature.

8.1. This is a dreadful time, rendered the more dreadful by the gloom of the weather and the country. I was never warm; my teeth chattered in my head; I was troubled with a very sore throat, such as I had on the isle; I had a painful stitch in my side, which never left me; and when I slept in my wet bed, with the rain beating above and the mud oozing below me, it was to live over again in fancy the worst part of my adventures — to see the tower of Shaws lit by lightning, Ransome carried below on the men's backs, Shaun dying on the roundhouse floor, a Colin Campbell grasping at the bosom of his coat. From such broken slumbers, I would be aroused in the gloaming; to sit up in the same puddle where I had slept, and sup cold drammach; the rain driving sharp in my face or running down my back in icy tricklets; the mist enfolding us like as in a gloomy chamber — or, perhaps, if the wind blew, falling suddenly apart and showing us the gulf of some dark valley where the streams were crying aloud. (Taken from *Kidnapped*, by Robert Louis Stevenson.)

8.2. On the various floors of this tower were the servants' quarters, and at the bottom a stable for horses. One room belonged to a woman who often looked after the younger children of the master of the house. At this particular moment Thomas was there, together with the little sister whose name we do not know. Suddenly there was a terrible thunderstorm and the tower was struck by lightning. The little girl was killed, and also the horses in the stable below. The mother in a state of panic rushed to the scene but found Thomas quite safe and sound, and still fast asleep next to the nurse. (Taken from *St. Thomas Aquinas*, by Angelus Walz, O.P., as translated by Sebastian Bullough, O.P. [Westminster, Md.: The Newman Press, 1951])

9. For each italicized predicate name the category and the predicable classification:

	Category	Predicable
9.1. Mary is *able to laugh*.	( )	( )
9.2. Mary is *laughing*.	( )	( )
9.3. Peter is *sentient*.	( )	( )
9.4. Peter is *six feet tall*.	( )	( )
9.5. Peter is *rational*.	( )	( )
9.6. Peter is *stoop-shouldered*.	( )	( )
9.7. Mary is *studying her lessons*.	( )	( )
9.8. Mary is *combing her hair*.	( )	( )
9.9. Mary is *lying down*.	( )	( )
9.10. This piece of paper is *white*.	( )	( )
9.11. Every organism is *living*.	( )	( )
9.12. Peter is *dressed in a white shirt*.	( )	( )
9.13. Peter is *taller than John*.	( )	( )
9.14. Peter is a *mathematician*.	( )	( )

THE CATEGORIES   71

9.15. Mathematics is a *science*.   ( ) ( )
9.16. Mary is *looking at the sunset*.   ( ) ( )
9.17. The sunset is *being admired by Mary*.   ( ) ( )
9.18. This tulip is *yellow*.   ( ) ( )
9.19. This tulip is a *plant*.   ( ) ( )
9.20. Mary is *nearsighted*.   ( ) ( )
9.21. Mary is *my sister*.   ( ) ( )
9.22. Paul was *hit by a bus*.   ( ) ( )
9.23. Paul was *thinking about his future*.   ( ) ( )
9.24. The missionaries were *enjoyed by the
       cannibals*.   ( ) ( )
9.25. Your sister is *in the next room*.   ( ) ( )
9.26. Paul is *courageous*.   ( ) ( )
9.27. The will is an *appetite*.   ( ) ( )
9.28. Red is a *color*.   ( ) ( )
9.29. Red is *striking*.   ( ) ( )
9.30. Paul is *armed to the teeth*.   ( ) ( )
9.31. Every body is *corruptible*.   ( ) ( )
9.32. The student was *worried about exams*.   ( ) ( )
9.33. Every maple tree is a *substance*.   ( ) ( )
9.34. The highway was *four lanes wide*.   ( ) ( )
9.35. Paul was *in Boston in February*.   ( ) ( )

10. Why can the contradictory of a term be used wherever its contrary or privative or relative opposite might be used?

11. Name the type of opposition involved between:
    11.1. ...... Double — half
    11.2. ...... Good — bad
    11.3. ...... Black — white
    11.4. ...... Teacher — student
    11.5. ...... Before — after
    11.6. ...... Logical — illogical
    11.7. ...... Virtue — vice
    11.8. ...... Intentional — nonintentional
    11.9. ...... Hearing — deafness
    11.10. ...... Sweet — bitter
    11.11. ...... Intentional — unintentional
    11.12. ...... Legible — illegible
    11.13. ...... Healthy — unhealthy
    11.14. ...... Sane — insane
    11.15. ...... Better — worse
    11.16. ...... Literate — illiterate
    11.17. ...... Love — hate
    11.18. ...... Dry — wet
    11.19. ...... Friendly — unfriendly

CHAPTER VII

Definition

I. Definition and Division

Simple apprehension admits of a more or less perfect grasp of its object. This object, as we have seen, is fully understood only in terms of its comprehension and extension. If there is only a vague, implicit knowledge of an object's comprehension and extension, then that object is known only obscurely or confusedly. This would be the case were I to know *animal* without knowing that its character as being *sentient* is intelligibly distinct from its character as being an *organism* or without knowing that *man* and *brute* are distinctly different immediate inferiors of which *animal* can be predicated. The act of simple apprehension is ordered to a much more perfect grasp of the objective concept than that given in this obscure way. Simple apprehension seeks the clear and distinct knowledge of objective concepts achieved by an explicit grasp of their comprehensive notes and extensive parts. Here we must note, however, that perfectly clear and distinct knowledge is difficult to achieve for any concept and, at least in practice, is impossible for most concepts. Yet we must try to realize such perfect knowledge as closely as possible, for scientific discourse is possible only to the extent that such knowledge is approached. The act whereby the intellect explicitly expresses the comprehension of a concept is the act of *defining* that concept. The act whereby the intellect expresses explicitly the distribution of a concept into its subjective parts is the act of *logically dividing* that concept. We shall use the terms "definition" and "division" to signify not only the acts of defining and dividing but also the objects known in each instance, as well as the terms used to express these objects. *Sentient organism* is the definition of *animal*; and *man* or *brute* is its logical division.

Definition and division are both instruments for manifesting the object of simple apprehension. Definition helps to disclose the meaning of an object. From this point of view the act of defining involves a movement of the mind. This means that the mind truly advances when it defines. However, from another point of view definition does not imply an advance in knowledge. The definition, as an object considered in itself, is identically the same as the thing defined and, as such, must be strictly coextensive with it. Thus, there is no movement from thing to thing involved in the act of defining. Yet the definition does differ conceptually from the defined, for it explicitly discloses the thought content which was only implicitly in the defined prior to its definition. Thus, in defining, there is an advance in knowledge from the point of view of the manner in which the object of simple apprehension is known. In the strict sense, the term *discourse* applies to the movement of the mind in the act of reasoning. However, in an analogical sense both the acts of defining and of logically dividing can also be spoken of as discursive.

Definition and division both presuppose the doctrine of the predicables and the categories. As we move through this chapter and the next we shall see how much the acts of defining and dividing depend upon the ability to distinguish between different predicables and different categories.

II. Nominal and Real Definitions

What we have said thus far about definition refers to the strictest type of definition, namely, that which reveals the meaning of a nature by making explicit reference to its comprehensive notes. Such definitions are rare, and in most instances we must be satisfied with less perfect, but attainable, definitions. The essential definition, as the definition which fully manifests the comprehension of the thing defined is called, is the primary instance of definition; and other definitions are more or less perfect to the degree in which they approach or fall away from this ideal. We shall begin to differentiate types of definition by distinguishing between *nominal* and *real definitions*. The essential definition, as we shall see, is a type of real definition.

As the names would indicate, a nominal definition expresses what a term stands for, and a real definition expresses what a *nature* is. This distinction, apparently easy to grasp, is unfortunately not so simple as it seems. Although we can give some characteristic examples of nominal and real definitions, there are many instances when the nominal and real definitions seem to be the same. *Science of man* would ordinarily be suggested as an example of a nominal definition for "anthropology." However, one might argue that this is as good a real definition of anthropology as possible. *Rational animal* is one of our clearest examples of a real definition. Yet in a given case it could serve just as well as a nominal definition. Hence it is worthwhile to consider the distinction between nominal and real definition more closely in order to clear up possible difficulties.

Confronted for the first time with a new term, e.g., early in the logic class with the term "simple apprehension" one asks, "What does this term stand for?" The answer would be that it stands for *the first operation of the intellect*. This is a nominal definition proposed in answer to a question asking for the meaning of a term. Now that the questioner knows precisely what thing is signified by the term "simple apprehension," he is ready to consider what this thing is in itself. He turns his attention from the term to the nature signified and asks, "But what is the nature of simple apprehension?" or "What is the first operation of the intellect?" The answer to this must be a real definition more or less perfect as it is closer or farther removed from a full and explicit disclosure of the comprehension of the nature concerned. The real definition of *simple apprehension* is *the operation of the intellect by which we grasp an object without affirming or denying anything about it*.

Hence a definition is nominal when it expresses the meaning of a term, and it is real when it manifests the nature signified by that term. The importance of language for the communication of knowledge is undeniable. Thus nominal definitions, which clarify the meanings of words, are highly important. But, as signs, words are important only in reference to the concepts they signify. Thus real definitions, which show what the nature is which these terms signify, are more important than nominal definitions. Nominal definitions are important in so far as they *point to objects known*. One cannot attempt to find out what something is (i.e., to discover, if possible, the constituents of its essence) until he first knows what it is he is to concern himself with. Whenever a term is so indeterminate as to leave this in doubt, a nominal definition is called for. Just as terms are signs of natures known, so definitions of terms (nominal definitions) prepare the way for definitions of natures (real definitions).

The search for nominal definition is a search for expressions which point to the things signified by the terms we are seeking to define. Nominal definitions are formed very frequently simply by reflecting upon the term to be defined. For example, the etymology of a term may reveal its meaning. "Philosophy" comes from the Greek word "philos" (lover or friend) and "sophia" (wisdom); hence it can be said to mean *the love of wisdom*. Similarly "psychology" comes from "psyche" (soul) and "logos" (discourse), and can be defined as *discourse on, or science of, the soul*. Synonyms of a term may also be used to define it, as well as equivalent expressions from another language. Thus "nuptials" can be defined as *wedding ceremonies* and "gendarme" as *policeman*. The nominal definition must be better known than the term it defines so that it will clarify the meaning of this term and point out precisely what the thing is that the term signifies. Thus an expression offered as (and suitably serviceable for) a nominal definition might also be a good example of a real definition. One might ask what "man" stands for and be satisfied with the definition *rational animal*. This could suffice, for example, to indicate that "man" was used to stand for the species *man* and not for a subclass of that species including only its masculine members. This same definition, offered as an expression of what the nature *man* is, is a highly acceptable real definition. Aristotle defines "soul" as *the first principle of life in the living being*. In so far as this expresses the meaning of the term "soul," this is a nominal definition. It could also be offered to express the nature of the soul, and thus it would be a real definition. Etymological definitions and definitions by way of synonyms and translation are rarely suitable as real definitions. But even here the name may be so appropriate that its etymology adequately defines the nature it signifies. Thus, as we have noted, "anthropology" can be nominally defined as *the science of man*, and this same expression can also be used as a real definition of *anthropology*.

III. The Four Causes

Before considering the different types of real definition it will be necessary to say something about causality and the four different kinds of causes. Real definition is frequently definition by way of cause, and some real definitions differ from one another because they express different causes. A cause is something from whose existence the existence of another follows. Cause and effect are correlative notions. A cause is some positive influence upon the being of something which follows from it. That which follows from the cause is its effect. An effect is an effect precisely in so far as it depends on its cause either to be what it is or to become what it becomes. Every corporeal thing depends for its being upon causes which fit into four types, namely, *material cause, formal cause, efficient cause,* and *final cause*.

Any corporeal thing is a certain kind of thing which can be changed into some other kind of thing. A green apple is a determinate type of apple which can be changed into a red apple through the process of ripening. Any apple whatsoever is a certain kind of substance which can be changed into man when eaten by a man and subjected to the process of nutritional assimilation. These examples differ, of course, in that the first has reference to what a thing is and can be accidentally, whereas the other refers to what it is and can be substantially. In both, however, we note that something is determinately a certain kind of being and can become a determinately different kind of being. A cause is required within a being to make it precisely the kind of being it is. This cause is the form of that being, its formal cause. The formal cause is determinate and determining. What it determines is an indeterminate and determinable substratum, which accounts for the ability of any corporeal being to change. This substratum, subject to one form at a given moment and with this form composing a certain kind of being, is able to give up this form and receive another, with which it composes a different kind of being. This indeterminate and determinable substratum, which is truly the subject of change, is called matter or the material cause. In so far as matter and form together constitute the essence of the being of which they are causes, they are spoken of as *intrinsic* causes.

The intrinsic causes of a corporeal thing do not fully explain the being of a corporeal thing. Matter, in itself indeterminate and lacking form, is unable of itself to supply the form which determines it in any given thing. There must be some agent, somehow extrinsic to both matter and form, which brings about the union of matter and form in any corporeal thing. This agent is called the efficient cause. But no agent operates as an agent without sufficient reason. The reason why an agent brings about the union of matter and form is the end or final cause. The efficient and final causes are *extrinsic causes*. Matter exercises its causality only in so far as it is actuated by a form. A form informs matter only in so far as it is made present to the matter through the efficient causality of an agent. An agent operates only in virtue of some final cause as motive. Thus the extrinsic causes are themselves related to the intrinsic causes as cause to effect.

Briefly, the material cause is the "stuff" out of which a thing is made. The formal cause is the design which is given the matter and determines the thing to be the kind of thing it is. The efficient cause is the agent which produces the thing. The final cause is the purpose or end of the thing. This does not mean that there must be only one material cause, one formal cause, one efficient cause, and one final cause for a given thing. The four causes are orders of causality, and any one corporeal thing may be the effect of a plurality of causes in each order. The doctrine of the four causes is of primary importance in understanding natural beings; however, it is perhaps most clearly illustrated for beginners in terms of artificial things. Thus a house is built out of bricks and wood (material cause) into a particular type of building (formal cause) by a corps of carpenters and bricklayers (efficient cause) so that it might serve as a protection against the elements (final cause).

This brief discussion of the orders of causality is not intended as an adequate philosophical exposition, yet it should serve as sufficient basis for a division of real definition according to difference in the causes involved.

IV. Types of Real Definition

The difference between real and nominal definitions is that the former reveal the meaning of a nature expressed by a term, whereas the latter simply point to what the term stands for. In the strict sense a nature is fully revealed only by its intrinsic causes, which are the causes which constitute its essence. In the case of corporeal beings (both natural and artificial) the intrinsic causes are the material and formal causes. Mathematical entities and incorporeal substances are, for different reasons in each case, essentially explained in terms of form alone. Whatever the kind of nature to be defined, the most perfect type of real definition is a definition in terms of intrinsic causality. This type of real definition is spoken of as the essential definition.

Because of the extreme difficulty and, in many instances, the impossibility of determining fully the intrinsic constitution of a nature, we are often unable to achieve essential definition. In such cases light may still be shed indirectly upon the essence in question by an appeal to something extrinsic to that essence, but *necessarily* connected with it. Two possibilities, and only two, suggest themselves here. First, it is possible to learn something about the nature by examining the extrinsic efficient and final causes, since these are causes in reference to the matter and form which constitute the essence in question. Thus, we might define a nature in terms of its extrinsic causes. Second, the attributes extrinsic to the essence but necessarily caused by it can indirectly show what it is. Thus, we might define a nature in terms of its properties. Definitions manifesting a nature in terms of its extrinsic causes or in terms of its properties are spoken of as nonessential definitions. A nonessential definition will be either a nonessential definition by way of efficient cause, a nonessential definition by way of final cause, or a nonessential definition by way of property.

Our investigation into the nature and types of definition thus far has yielded results which we can present schematically as follows:

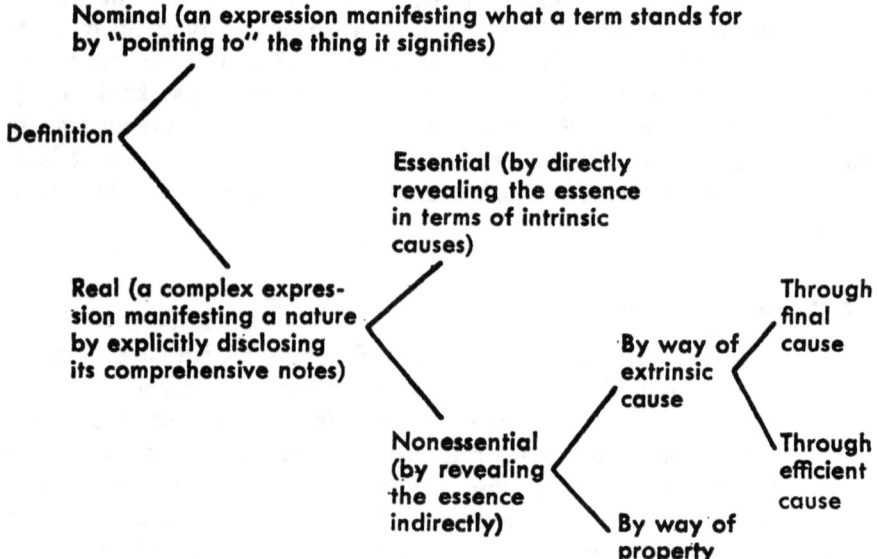

We have discussed the difference between nominal and real definitions at some length. Now we shall look more closely at the types of real definition.

Essential Definition

An essential definition manifests a nature by expressing its intrinsic causes. These can be expressed as entitatively, i.e., really, and not just intelligibly distinct parts joined together in the composite they constitute, which is, of course, the nature in question. (For immaterial entities the form alone would be expressed, not as a part but as the whole of the essence to be defined.) For example *man can be essentially defined as a being composed of a rational soul and a natural body potentially having rational life within it.* The parts of this definition are respectively the form and the matter of human nature, which together constitute the essence of man. But the better way of expressing an essential definition logically is to do so in terms of the proximate genus and the specific difference of the defined. A definition in terms of proximate genus and specific difference is an essential definition (i.e., a definition through intrinsic causes), for the genus is taken from the material cause and the difference from the formal cause. This second way of expressing the essential definition is logically more strict than the first because it is more suited to logical analysis and scientific discourse. It is easier to see how the definition expressed in this second way is logically related to other intelligible objects than the definition expressed in the first way.

Definition by Way of Extrinsic Cause

Since both the final and the efficient causes of a thing exert a definite influence in shaping its essence, it is possible to manifest the essence in terms of final and/or efficient causes. Here, however, the manifestation is indirect, whereas it is direct in the case of definition by intrinsic causes. Thus, when an essential definition is impossible or when a complementary definition is desired, definitions by way of extrinsic causes may be called for.

Nonessential definitions are naturally less perfect than essential definitions; yet good nonessential definitions should approach essential definitions as closely as possible. Ordinarily it is difficult, and frequently impossible, to determine the specific difference which is necessary for an essential definition. If it is difficult to determine proximate genera, it is usually possible to find the remote genera. The crucial problem in attempting a nonessential definition is to determine an adequate substitute for a specific difference. Here we should recall why a *specific difference* is so important for an essential definition. The reason is that the specific difference pinpoints the definition so that it applies only to the thing defined, distinguishing it from all other things. When we cannot discover a specific difference and try to substitute for it some extrinsic cause or causes, we are attempting definition by way of extrinsic cause. To do this, the extrinsic cause or causes must contract the genus precisely to the thing defined. Man can be defined as an *animal destined for eternal life*. This is a good nonessential definition through final cause, for *animal* is the proximate genus of *man*, whereas *destined for eternal life* expresses a final cause which is limited as far as animals are concerned to the species *man*. Certain things are characteristically defined through final and/or efficient causes. For example, artificial things, especially tools and other useful objects, can ordinarily be defined clearly only in terms of the purpose for which they are intended. And many diseases are frequently best defined in terms of the bacteria, germs, or other agents which cause them.

Definition by Way of Property

Attributes flowing from an essence are extrinsic to it; however, at the same time they reveal something about that essence in so far as they are caused by it. Thus these attributes can be used in nonessential definitions. Since the definition and the thing defined must have the same extension, only attributes related as properties to an essence can be used to define it. A definition in terms of an accidental (rather than proper) attribute would be either too broad or too narrow in extension for the concept it was intended to define. Only an attribute that is necessarily and exclusively an effect of the specific difference of the species can substitute for that specific difference in a nonessential definition by way of caused attribute. Thus, this type of nonessential definition is called definition by way of property. *Man* can be defined in this way as *an animal endowed with speech*, or as *a tool-bearing animal*, or as *an artistic animal*. *Triangle* can be defined as *a plane figure with three angles equal to two right angles*. *Body* can be defined as *a corruptible substance*.

We have seen that accidents cannot substitute for a missing specific difference in a legitimate nonessential definition. This is true for any individual accident. However, it is possible that a combination of accidents might function as a property for a given thing. So long as a number of accidents understood collectively belong necessarily and always to every member of a species and only to the members of that species, this combination of accidents has the force of a property and can be used in a definition by way of property. However, it is extremely difficult to understand any combination of accidents well enough to see in their combination the necessity of a strict property. Some have suggested that man can be defined as a *two-legged, featherless, nonaquatic animal*. *Two-legged* distinguishes all actually existing men from all known animals other than men, with the exception of birds. *Featherless* distinguishes men from all known birds except penguins. *Nonaquatic* distinguishes men even from penguins. This combination of accidents does serve to distinguish men from all other things in our experience. However, unless we can see a conceptual incompatibility between this combination of accidents and *nonman*, and the lack of this combination of accidents and *man*, then this definition falls short of being a legitimate real definition. It can, however, surely serve as an acceptable nominal definition of "man"; for it does point among the things within our experience precisely to man, thus isolating *man* accurately as an object for inquiry.

V. The Rules for Good Definition

Actually a set of rules for good definition is not necessary for one who understands the nature of definition. This is true because such rules do no more than repeat, after the fashion of expressed norms, the characteristics proper to definition. However, such rules are useful for beginning students, as they remind them of the requirements of a good definition. Since the primary instance of definition is essential definition, the rules apply primarily to it. They apply proportionately to lesser types of definition.

1. Definitions should include generic and differentiating elements. The former shows what it has in common with other things, and the latter shows how it differs from them.

Clearly, in the strictest type of essential definition, the generic element is the proximate genus and the differentiating element is the specific difference. Other definitions, with the exception sometimes of the nominal definition, should contain a generic element as close to the proximate genus as possible and a differentiating element which can substitute adequately for a missing specific difference. Neither analogical concepts nor ultimate genera can be defined in terms of proximate genus and specific difference, for neither has a genus or specific difference in the strict sense. The best that can be done here is to approximate an essential definition by way of elements like, but not identical with, proximate genus and specific difference. The definition of an accident also presents a special case. Essentially an accidental form, e.g., the ability to laugh, is purely a form without matter. An accident, however, can be understood only in reference to the substance which is its proper subject. Hence the two elements in the definition of an accident will express not only the form which is the accident but the matter which is its proper subject, e.g., *man* as proper subject of *ability to laugh*.

2. The definition and the defined should be convertible (neither one more or less extended than the other).

A definition is intended to clarify as far as is possible the thing it defines. Thus, it must apply to every instance of what it defines and only to what it defines. Anything less than this will confuse rather than clarify. If *triangle* were defined as *a three-sided plane figure with equal sides* one would think that only equilateral triangles, and not scalene or isosceles triangles, were triangles. If *triangle* were defined as *a rectilinear plane figure* one would confuse triangles with quadrilateral figures and with every other type of polygon. The only adequate essential definition for triangle is *three-sided plane figure*. Every triangle is a three-sided plane figure, and every three-sided plane figure is a triangle.

3. The definition should be clearer than the defined.

As noted, the purpose of a definition is to clarify the thing it defines. It is obvious, therefore, that the definition must be clearer than what it defines. This means that the definition ought to be as brief as possible. It should not be in metaphorical terms. It should be in terms which can be understood by those for whom the definition is intended. It should be in positive rather than negative terms whenever this is possible. This rule demands, especially, that the definition should not contain the defined or any expression which is a derivative of the defined. Of course, the thing defined must be known, indistinctly, prior to its definition and before the parts of its definition are known precisely as such. But these parts must be understood themselves before the definition as a whole is understood. Thus, *man* is known before *rational* and *animal* are known as parts of its definition. But *rational* and *animal* must themselves be understood before they can together manifest the signification of *man*.

4. The defined must be universal.

The ultimate intention of definition is to express distinctly the comprehension of a nature which is known in an act of simple apprehension, i.e., of a universal concept. Thus, definitions are possible only for universals. Definitions are unchanging and eternal, because they express the meaning of essences which have been abstracted from changing individual situations. Individuals can be identified by descriptions involving the definitions of their species, but the ultimate difference pinpointing an individual will have to be some singular characteristic of contingent and perhaps even momentary status.

Exercise VII

1. Define: definition, logical division, logical discourse, nominal definition, real definition, cause, material cause, formal cause, efficient cause, final cause, intrinsic cause, extrinsic cause, essential definition, nonessential definition, definition by way of extrinsic cause, definition by way of property.
2. In what sense is definition an instance of logical discourse, and in what sense does it fall short of discourse strictly taken?
3. Explain: definition is to comprehension as logical division is to extension.
4. How do nominal definitions differ from real definitions? How do nominal definitions serve real definitions?
5. Define each of the four orders of causality, and give examples for each order.
6. Explain the manner in which each cause is related to the others.
7. How does the essential definition differ from the nonessential? What are the two different ways of expressing an essential definition? Which is the stricter way from the point of view of logic? Why?
8. What are the two types of nonessential definition? Why are there these two types, and only these? Subdivide these types where this is necessary, and illustrate each type of definition.
9. When can an accident (rather than a property) be used in a real definition? What are the strict limits to this use?

DEFINITION

10. List the rules for a good definition, and defend each rule by measuring each one against the definition of definition itself.
11. Classify each of the following definitions which are good, and explain what is wrong with each of the bad ones. For each good real definition point out the generic element and the differentiating element.
    11.1. Man — a creature destined for eternal life.
    11.2. Science — an organized body of knowledge.
    11.3. Triangle — a plane figure bounded by three lines.
    11.4. Osmosis — the passage of water through a membrane.
    11.5. Second intention — a logical property belonging to an objective concept precisely as known.
    11.6. Edward — rational animal.
    11.7. Zygote — fertilized egg.
    11.8. Logic — the art of proceeding logically.
    11.9. Definition — a principle of manifestation for the object of simple apprehension.
    11.10. Indian chief — top man on the totem pole.
    11.11. Antibiotic — a chemical agent possessing bactericidal properties.
    11.12. Human soul — principle of life.
    11.13. Intelligence — the ability to solve problems.
    11.14. Man — tool-bearing animal.
    11.15. War — hell.
    11.16. Quality — that kind of a being apt to exist in another and not in itself.
    11.17. Bicycle — a light vehicle having two wheels, one behind the other, a steering handle, a saddle seat or seats, and pedals by which it is propelled.
    11.18. Sign — that which signifies something beyond itself.
    11.19. Substance — that kind of being apt to support accidents.
    11.20. Man — creature composed of body and soul, created by God, to give glory to God and share His happiness forever.
    11.21. Organism — living body.
    11.22. Body — corruptible substance.
    11.23. Mathematics — the science of numbers.
    11.24. Telescope — an optical instrument used to aid the eye in viewing distant objects.
    11.25. Law — a reasonable ordinance, promulgated by the one in charge of the community, for the sake of the common good.
    11.26. Politics — a branch of ethics specifically distinct from other branches of ethics.
    11.27. Poker — a gambling game played with cards.
    11.28. Gland — a cell or collection of cells which produces a determinate product.
    11.29. Accident — nonessential predicable.
    11.30. Jazz — music which is improvised spontaneously.
12. Why is it so difficult to define the several categories?
13. Check this textbook for definitions. Classify and evaluate them chapter by chapter. Do the same with other books. Prepare a list of faulty definitions found in both technical and nontechnical works.

CHAPTER VIII

Division

I. Logical Division

A definition, as we have seen, is a complex expression explicitly stating the comprehension of a concept. Logical division, on the other hand, looks to extension rather than to comprehension. It is thus a complex expression which manifests the extension of a concept by pointing out what subjects, i.e., inferiors, are included within that extension. Just as the comprehension of *animal* is made explicit in the definition *sentient organism*, so the extension of *animal* is made explicit in the logical division *man or brute*. The act of defining is an intellectual operation which clearly and distinctly manifests the actual content, or meaning, of a concept. The act of logically dividing is an intellectual operation which clearly and distinctly discloses the potential content of a concept. That is, it shows how this concept can be realized in various types of subjects. Each renders the relatively unknown — because only implicitly grasped in an obscure and confused fashion — known. As such, each is a kind of discourse, although we shall use the term "discourse" more strictly later on in reference to the more complicated process of reasoning. Just as essential definition answers the question, "What is this nature?" so logical division answers the question, "In what different kinds of beings can the nature be realized?"

Every logical division involves three elements. These are: the logical whole or totality divided, the subjective parts or dividing members, and the principle of division or aspect of the whole in virtue of which it is divided into parts. When *animal* is divided into *man or brute, animal* is the totality divided; *man* and *brute* are the dividing members; and specific differentiation is the principle of division.

For division as well as for definition the logician is concerned only with the general theory. He defines division and divides and subdivides it into the general types, and then defines those types. However, he leaves the particular application of the theory of division for the proper subjects of the special sciences to the scientists in question. The particular method of division best suited to the subject matter of a given science can be determined only by one who is familiar with that subject matter.

II. Types of Logical Division

Just as the primary instance of definition is essential definition, so the primary instance of logical division is essential division. This is the division of a genus into its species through the addition of specific differences. If the basis for logical division is something other than specific differences, then the division is nonessential rather than essential. The division of *animal* into *man* or *brute* is an essential division because it is made on the basis of diverse specific differences. The divisions of *animal* into *male* or *female* and of *man* into *sane* or *insane* are nonessential divisions. In all of these cases, the differences which are the basis for the divisions do not belong to the essence of the nature divided. They are not in the same category as that nature and the members into which the latter is divided are not related to it as species to genus.

All the different nonessential aspects under which a universal concept can be said to be realized in different types of beings can serve as the basis for nonessential divisions. Traditionally the modes of nonessential division have been classified under three headings: (1) the division of substances by way of their accidents, (2) the division of accidents by way of the substances in which they inhere, and (3) the division of accidents themselves by way of other accidents. *Man* is divided into *white, brown, red,* or *yellow* according to the first mode. *Able to reproduce one's kind* is divided into *animal* or *plant* according to the second. *What is cold* is divided into *liquid, gas,* or *solid* according to the third. All that is needed for a legitimate nonessential division is that the whole must be divisible into parts on the basis of mutually exclusive or opposed differences which do not belong to the essence of the whole. To illustrate the variety of possibilities, let us note several nonessential divisions, each involving a different mode of opposition. The division of *disease* into *crippling* or *noncrippling* is based on contradictory opposition between differences. The division of *number* into *odd* or *even* is based on immediate contrariety. The division of *man* into *logical* or *illogical* is based on privative opposition. Finally, the division of *soldier* into *superior* or *inferior* rests on the opposition of relation.

Analogical concepts pose a problem for logical division just as they do for definition. They cannot be defined as strictly as univocal concepts because they do not admit of genus and difference in any strict sense. And they cannot be logically divided in the same way as univocal concepts, namely, by the *addition* of opposed differences to which they are in potency. The differences between the inferiors of an analogical whole are actually present in the whole, although their presence together is not distinctly known. Here the inferiors are differentiated from one another as differing modes of the analogical whole. This is the case when *being* is divided into *substance* and the *nine accidents*, as we saw previously in our chapter on the categories.

Logical *division* can itself be divided into *dichotomous division* and *nondichotomous division*. A dichotomy is a division in which two dividing members exhaust the totality divided. A dichotomy is the simplest possible division. A dichotomous division is the only kind of division possible whenever there are but two subjective parts of the whole from the point of view chosen for division. The division of *whole number* into *odd* or *even* is a legitimate dichotomous division which allows for no alternative, since the principle which distinguishes some whole numbers as odd necessarily distinguishes all others as even. The division of *triangle* into *equilateral* or *nonequilateral* is legitimate, but it is not the only division of triangle which can be made from the point of view of the relations between its sides. Nonequilateral is a negative notion which, positively expressed, would include both *scalene* and *isosceles*. Under most circumstances the division of *triangle* into *equilateral*, *scalene*, or *isosceles* would be preferred to the dichotomy including *equilateral* and *nonequilateral*. However, if the intention of the division is ultimately to focus attention on only one type of triangle, namely, the equilateral, the dichotomous division may, because of its simplicity, be more helpful. Suppose, for the moment, that in dividing *triangle* one knew there were equilateral triangles and other kinds as well. Suppose, however, that he did not know precisely what these others were. In such a case, the dichotomy between *equilateral* vs. *nonequilateral* would be safe, whereas an attempt to divide *triangle* into *equilateral* and more specific types might lead to error. The division of *triangle* into *equilateral* or *nonequilateral*, as of *logical division* itself into *dichotomous* or *nondichotomous*, is a division by way of contradiction. Division by way of contradiction is frequently helpful, and it is usually safe. However it has one danger. The divisions of man into *white* or *nonwhite*, *Asiatic* or *non-Asiatic*, *rich* or *nonrich*, are safe and legitimate, although they have only slight value. However, the division of *man* into *risible* or *nonrisible* is both useless and illegitimate, although it might seem at first glance to be valid. The difficulty, of course, is that all men are risible. There can be no subjects in whom the second member of this attempted dichotomous division can be realized. Nonrisible simply cannot be a subjective part of man, and never, then, a member of any logical division of man. This means that dichotomous division by way of contradiction cannot be attempted if one of the members is equal to the totality which is being divided.

III. Physical Division

Logical division is defined as the orderly distribution of a logical whole into its subjective parts. We can extend the notion of division to include both logical division and a second kind of division which we can call physical division. Generally taken, division is defined simply as the distribution of any whole into its parts. Logical division distributes the logical whole into its subjective parts. Physical division divides an integral whole into its composing parts. A logical whole necessarily involves subjective parts, that is, parts which are included within its extension and which together make up the whole of its extension. Man and brute together make up the whole of the extension of animal. They are the subjective parts of animal, and animal is a logical whole in reference to them. Thus animal can be said of each of them. An integral whole is actually the sum of its parts. Collectively they make it actually what it is, and are its composing parts. The human body is an integral whole actually composed of head, limbs, trunk, and extremities. No one of them is the body, but together they compose the body. Unlike any logical whole, every integral whole actually and distinctly includes its parts and is, in respect to its actual content, the sum of these parts. Unlike any logical whole, an integral whole cannot be said of any one of its parts. Every man is an animal, and every brute is an animal. But no head is the body, nor is any limb or any other composing part of the body able to be said to be the body.

There are many different kinds of integral wholes in the broad sense in which we have used this term. We have seen that the distinct parts of any body are composing parts of an integral whole. The entitative parts of any nature are also composing parts of an integral whole. So too are the generic and differentiating elements in a definition or the several comprehensive notes of an objective concept. What is common to each of these instances of the integral whole and its composing parts is this: the whole involved is actually the sum of its parts and cannot be said of any of them. There are as many examples of physical division as there are integral wholes which can be divided into composing parts. The division of the United States of America into fifty states is a physical division. The division of human nature or of any individual man into body and soul is a physical division. The division of the definition *rational animal* into *rational* and *animal*, and of the concept *man* into *substance, material, living, sentient,* and *rational* are also physical divisions. Notice that logical division requires a universal whole, whereas the whole required in physical division may be either universal or singular.

IV. Rules for Good Division

The rules for good division are but a reaffirmation of the nature of division itself. They apply both to logical division and to physical division.

1. The dividing members must be related to the totality divided as parts to whole (i.e., in the case of logical division, as inferiors to superior; and in the case of physical division, as composing parts to an integral whole).

This rule is an immediate consequence of the definition of division and needs no defense or explanation. This rule is violated by any division in which one of the dividing members is coextensive with the totality divided (e.g., *man* divided into *risible* or *nonrisible*).

2. The dividing members must adequately divide (i.e., exhaust) the totality divided.

The purpose of division is to give a clear and distinct understanding of a whole in terms of the parts which make it either potentially (in logical division) or actually (in physical division) what it is. If the division included only some of the parts comprising the whole this would hinder rather than help in the understanding of the totality divided. The division of *predicable* into *species, genus, difference,* or *property* suffers this defect. *Accident* is required to complete the list of predicables and render it adequate as a legitimate division. Every universal concept involves an infinite multitude of singulars in its extension. Hence, there can be, strictly speaking, no division of a logical whole into its singular inferiors. The attempt to list singulars under a logical whole is an enumeration rather than a division properly understood.

3. The dividing members must be formally opposed to one another, so that each part excludes the other.

A good division requires that the dividing members must be clearly differentiated from one another. None of the parts should include another either in whole or in part nor should any of the parts be identical. Any failure here renders the attempted division disorderly. To divide *football player* into *lineman, back,* or *tackle* is confusing and illegitimate, for *lineman* includes *tackle* as one of its types.

4. One and the same principle of division must be observed throughout a division.

In a sense the principle or basis of division is the formal element in division, for it makes the division to be precisely the kind of division it is. Multiply principles of division and the result is chaos; for then there is a division which is simultaneously more than one division. Failure to observe this rule ordinarily breaks the previous rule. Even when this is not the case, the division is meaningless and confusing. To divide *people* into *men, women,* or *children* is confusing or disordered, so long as it is based, simultaneously, on a difference in age and on a difference in sex. To divide *animal* into *risible* or *nonmoral* yields a set of dividing members which happen to exhaust the totality divided while mutually excluding one another. Yet the division is not clear as it stands, since one dividing member represents a part of the whole considered from one point of view and the other a part of the whole considered from an entirely different point of view. To make this division meaningful it would be necessary to reduce *risible* to its source in man's rationality and *nonmoral* to its source in the brute's nonrationality.

V. Codivision and Subdivision

In many cases there is no reason why division must cease once a totality has been divided into a set of parts. As a matter of fact, there may be compelling reasons to divide each of the original dividing members into its own parts and perhaps to continue this process through several levels of division. To divide a whole into its parts and to divide the parts of the original whole into their own parts is to subdivide. The division of *definition* into *nominal* or *real,* coupled with the further division of *real definition* into *essential* or *nonessential,* and finally of *nonessential real definition* into *definition by way of extrinsic cause* or *definition by way of property,* is an example of legitimate subdivision.

Sometimes the whole divided into one set of parts according to one principle of division can be divided into another set of parts from a different point of view. To divide the same totality several times, each time by a different principle of division, is to codivide. In the first chapter of this book we divided *logic* into *the logic of the first operation, the logic of the second operation,* and *the logic of the third operation;* and we divided it also into *formal logic* and *material logic.* This is an example of legitimate codivision.

The rules which apply generally for division apply, of course, for subdivision and codivision. Each distinct division included within the complex framework of either subdivision or codivision must measure up to the four rules discussed in the preceding section. The more complicated the process of subdivision or of codivision, the more carefully must we attend to the rules. Increasing complexity brings with it a correspondingly increasing possibility of disorder, and disorder is the death of division.

VI. The Value of Definition and Division

Both definition and division have an importance proper to themselves. Each aids in explaining the object grasped in the first operation of the intellect. Definition clarifies the concept from the point of view of comprehension, while logical division clarifies it from the point of view of extension. It is a philosophical axiom that act is prior to potency, in the sense that the actual is more excellent than the potential. Since comprehension expresses what the concept is in act, while extension expresses what it is in potency, definition is logically of greater importance than division. Accordingly, it is not surprising to note that division is of value not only in reference to itself, but also in so far as it is ordered to definition. The ultimate perfection of the first operation of the intellect comes in defining, and frequently definition can be achieved only in conjunction with division. For example, the dividing members in an essential division are differentiated from each other by opposed specific differences and are united in a common proximate genus. Essential divisions, therefore, attain differing species. The process of dividing gives us the elements needed for the essential definition of these species, namely, proximate genus and specific differences. The same is true, in proportionate fashion, for less strict instances of definition and division. Definition is a process which demands for its integrity a high degree of precision. Unless the defined and its definition are but different expressions of identically the same thing, the definition in question is faulty. Division is an extremely useful logical instrument for attaining the precision needed in definition. To illustrate: mathematics in the classical sense can be defined as the speculative science of quantified being. In attempting to understand this definition one might begin by noting that of the ten ultimate types of being mathematics fits into the category of quality. Then it is significant to note that of the four species of quality mathematics falls under the first as a habit. Habits are either entitative or operational, and mathematics is of the second type. Operational habits are either good or bad. Good habits are virtues, and mathematics is a virtue. Virtues in turn are either moral or intellectual, depending on whether they perfect the will or the intellect. Mathematics is an intellectual virtue, and that special kind of intellectual virtue which we call science. Science itself is either practical or speculative, depending on whether its end is action or truth. Mathematics aims at truth, and so is a speculative science. Speculative science, finally, is either ordered to a knowledge of physical being, quantified being, or being as such. Mathematics is the second type, namely speculative science of quantified being. Thus the definition of mathematics is arrived at, in this example at least, after a fairly complicated set of divisions and subdivisions. The way to definition need not always be as involved as this. The purpose of the example is not to manifest an unvarying process. It simply illustrates the connection between definition and division and the assistance that division can render in the approach to definition. The illustration also points up the fact that definitions are not always easily achieved. The formation of a definition is usually the result only of a long and complicated logical effort. Division is one of the chief instruments of the intellect in its search for definitions.

As long as the intellect remains on the level of simple apprehension, it is ordered to the act of defining. But the first operation of the intellect is itself ordered to the second and to the third. Thus, definition is not simply an end in itself. It is primarily of concern in so far as it can be a principle of argumentation. We shall return later to the question of definition in order to see how definition does function as a principle of argumentation.

Unmistakable evidence of the importance of both definition and division to science can be found in any scientific textbook, including this one. Each chapter abounds in definitions and divisions of one type or another. An examination of the two sections on definition and division in this book reveals a variety of definitions and divisions. For example, definition is logically divided and subdivided into its types, and each of the dividing members in this logical subdivision is itself defined. In addition, each of the dividing members of this logical subdivision is physically divided into its composing parts. Consider the case of definition by way of property. We saw that this is a type of nonessential definition. Nonessential definition is a type of real definition. Real definition is a type of definition. And definition is in a sense a type of logical discourse. We defined definition by way of property as a nonessential definition indirectly manifesting the essence of the defined in terms of a necessary and exclusive attribute caused by that essence. Finally, we physically divided it into the genus of the defined and one of the properties of the defined. Division too has been defined and divided (both logically and physically) with its dividing members themselves defined and sometimes further divided. This is as it must be. There can be no systematic and scientific discourse that dispenses with rigorous definitions and divisions. Not that these represent the whole of any scientific treatise nor the end of any scientific inquiry. But they do represent an indispensable part and a beginning without which there can be no significant scientific conclusions.

Exercise VIII

1. Define: logical division, essential division, nonessential division, totality divided, dividing members, principle of division, dichotomous division, physical division, logical whole, subjective parts, integral whole, composing parts, codivision, subdivision.
2. Compare logical division to definition. Why is each spoken of as discursive? What are limits to this manner of speaking?
3. Logically divide nonessential division into its types; explain the reason for the division in terms of the principle used for making it.
4. What are the modes of opposition which are significant for the theory of division?
5. What are the advantages and possibly some of the disadvantages of dichotomous division?
6. How precisely does a logical whole differ from an integral whole?
7. Define physical division and then illustrate the many ways in which physical division can be realized by supplying examples of significantly different physical divisions.
8. What does it mean to say that for logical division the whole is in the part, and for physical division the part is in the whole?
9. List the rules for a good division, and defend each rule by measuring it against the definition of division itself.
10. Compare subdivision with codivision, and discuss the rules for a good division in the context of each.
11. What is the value of division considered in itself? What is the value of division as ordered to definition?
12. Classify and evaluate the following attempts at division:
    12.1. Predicable — species, genus, generic difference, specific difference, property, and accident.
    12.2. Science — philosophical science and experimental science.
    12.3. Division — logical and physical.
    12.4. Organism — plant, animal, and man.
    12.5. Essential definition — proximate genus and specific difference.

DIVISION 93

12.6. Virtue — good habit, intellectual habit, and moral habit.
12.7. Man — artistic and nonartistic.
12.8. Student body — freshmen, sophomores, juniors, and seniors.
12.9. Student — freshman, sophomore, junior, and senior.
12.10. Snow — white and nonwhite.
12.11. Logic — logic of the first operation, logic of the second operation, and formal logic.
12.12. Digestive tract — oral cavity, pharynx, esophagus, stomach, small intestine, and large intestine.
12.13. Clothing — black and white.
12.14. Organism — living body, plant, and animal.
12.15. Predicate — essential predicable and nonessential predicable.

13. Logically divide, then subdivide and/or codivide the following:
    13.1. Quality
    13.2. Cognitive operation
    13.3. Football player
    13.4. Substance
    13.5. Law
    13.6. Society
    13.7. Predicable
    13.8. Logical discourse
    13.9. Plane figure
    13.10. Pleasure

14. Make a list of the divisions employed in this textbook, and in other technical and nontechnical books. Classify and evaluate each.

15. There is a lengthy and logically rigorous passage in St. Thomas' *Commentary on the De Anima of Aristotle* in which several divisions are proposed in preparation for an adequate definition of the soul. Study this passage and pick out each division and each definition offered in it. Classify and evaluate each one of them. Try to reproduce the passage in outline form. The passage in question is Book II, Lesson 1, No. 213–233 (Pirotta edition). It is available in English, translated by K. Foster and S. Humphries in *Aristotle's De Anima* (New Haven, Conn.: Yale University Press, 1954), pp. 166–172.

PART II

The Logic of the Second Operation

CHAPTER IX

Judgment: The Second Operation

I. The Nature of the Second Operation of the Intellect

The first operation of the intellect bears upon quiddity or essence — and nothing more. In simple apprehension we grasp natures without affirming or denying anything about these natures, not even their existence, actual or possible. Thus simple apprehension is a truncated act of knowing. It gives us at best an incomplete and unsatisfying view of reality. It can be called an analytic act which knows reality only by "tearing it apart" and presenting but one aspect of it. Knowing is completed only in an operation which "puts reality back together again" by grasping essences in the only way in which they are, namely, as exercising some type of existence. This is the second operation of the intellect, which we call judgment, whose expression is the proposition.

Judgment is the act of the intellect affirming or denying one thing of another. There are basically two types of judgment. These are the existential judgment and the attributive judgment. The existential judgment is the less complex of the two. It is the intellectual act of knowing that something is (i.e., exists) or is not (i.e., does not exist). The attributive judgment is more complex. It is the intellectual act of knowing that something exists or does not exist with certain determinations. *This man is* (i.e., *exists*) and *This man is not* (i.e., *does not exist*) are existential propositions, while *This man is tall* and *This man is not tall* are attributive propositions. In the one case existence itself is affirmed or denied of a subject. In the other case some quidditative determination is affirmed or denied of a subject. In both cases both quiddity and existence are involved: in the first *something* is thought *to be*, and in the second *something* is thought *to be something*. In each case something (which we shall call the logical subject) is thought *to be* somehow modified or actualized by something else (the predicate). The predicate is related to the subject as form to matter, because the predicate is understood to signify some sort of determination or perfection for the subject.

The second operation of the intellect is frequently referred to as the operation of composition or division. The reason for this is most clearly seen in the case of the attributive judgment. In the affirmative attributive judgment two intelligible objects are united or composed by way of affirmation. In the negative attributive judgment two intelligible objects are separated or divided by way of denial. *Man* and *able to talk* are two of the many distinct intelligible aspects existing together in subjects such as Mary, John, and Peter. They are distinct as intelligible aspects, and each is known in a distinct act of simple apprehension. Each has its own comprehension and is defined in its own way. In a word, they are distinctly different objects with distinctly different meanings. Yet, even though *man* and *able to talk* are different intelligible objects they exist together in the same thing, namely, this man, that man, and in fact every man. As intelligible objects they are two, but as realized in existing subjects they are one. To judge *Every man is able to talk* is to grasp *man* and *able to talk* in their existential status, knowing as one in existence things which are two as objects of thought. This is to know that *man* is identical in subject with the *being which is able to talk*. *Man* and *able to talk* remain diverse in meaning, and the affirmative attributive judgment in no way intends to identify them in meaning. It identifies them only in reference to existence by affirming them to be two aspects of the same thing.

Judgment bears on existence as well as essence. The affirmation or negation of the second operation is an intellectual commitment to the way in which things are. This is clearly the case whenever the subject about which something is affirmed or denied is an actually existing thing. But *Every man is able to talk* would be true even if no men actually existed, and *Santa Claus wears a beard* is true although Santa Claus is only a fiction. It seems perhaps that we can make a judgment in instances where the judgment need not or does not bear upon existence. However, this is not the case. There is no affirmation or denial unless there is some reference to the way things are, and there is no judgment apart from affirmation or denial. The primary instance of existence is the act of existing which makes a thing to be an actually existing subject outside the mind — the kind of existence enjoyed by each one of us right now. However, this is not the only type of existence. Things can exist in the mind, as objects known, as well as outside the mind in the realm of real subjects. Two types of things can exist in the mind: first of all, those things which also can exist outside the mind, and second, those things which cannot exist outside the mind precisely because extramental existence is contradictory to their natures. Those which can exist outside the mind can be divided into those which actually exist extramentally and those which do not but could. We have called the first actually real beings and the second possibly real beings. We have seen that those things which can exist only as objects in the mind are called beings-of-the-reason. The most significant type of judgment refers to beings which exercise the act of real existence. But judgments about possibly real beings and beings-of-the-reason are legitimate.

II. The Pre-eminence of the Second Operation

Judgment is the most important operation of the intellect. We have seen that simple apprehension is a truncated act of knowing, which is related to judgment as the imperfect to the relatively perfect. Reasoning presupposes prior judgments. Nevertheless reasoning is ordered to judgment. Reasoning itself makes sense only in so far as it generates a conclusion. This conclusion is itself grasped in an act of judging. Were we able to compose and divide in every instance without the support of premises, which are themselves compositions or divisions, we would never reason. Man is not equipped intellectually so that he can do this. He is by nature a rational or reasoning being. When he does reason he reasons so that he may judge. Thus, the first and third operations are each in their own way for the sake of the second. The second operation is the intellectual operation par excellence.

III. Judgment and the Motive for Assent

A judgment is an active intellectual commitment in the face of evidence deemed sufficient to warrant assent. Judgments can differ from one another either on the basis of the firmness of the assent or on the basis of the nature of the evidence for assent. As far as the firmness of assent is concerned, some judgments are certain and some are only probable. A judgment is certain when the assent given to a proposition excludes the fear that its opposite is true. This is the way in which I firmly assent to the truth that every man is rational. There is absolutely no reason to suspect that some man is not rational. A judgment is probable when assent is given to a proposition with an awareness that its opposite might possibly be true. Thus I may assent to the proposition that some runner will run a mile in three minutes and fifty seconds. However, I assent to this only with reservation, for I am aware of the possibility that no runner may achieve this mark. We are speaking of the difference between certain and probable judgments on the basis of a difference in firmness of assent. Lest this be misunderstood, we should note that even in judging with probability there is a sense in which a firm assent is given, but it is a firm assent only to the probability of the proposition in question. I am not sure that a runner will establish a record of three minutes and fifty seconds for the mile run, but I am sure that it is a distinct probability.

Judgments can also differ on the basis of the evidence which makes them legitimate. Assent to a proposition may be given on the basis of extrinsic or intrinsic evidence. Extrinsic evidence consists in the testimony of some authority to the truth of a proposition. This is assent by way of faith. Faith is natural or supernatural depending on whether the authority in question is finite or divine. Faith yields certitude or probability depending upon the degree of the trustworthiness of the authority involved.

Intrinsic evidence comes from the thing itself, not from some authority who knows about it. Such evidence may be either immediate or mediate. Immediate evidence is found in a direct sensory-intellectual experience of an actually existing thing or in the very meanings of the terms involved in a proposition. The first kind of immediately evident proposition is factually evident, the second self-evident. Propositions are mediately evident when they are seen to follow by way of reasoning from either type of immediate proposition. The certitude or probability of assent based on intrinsic evidence depends on the force of the evidence in question.

Consider the following examples which illustrate the differences in judgment considered in this section:

1. I know certainly on the basis of supernatural faith the fact that there is a final judgment for each man — because God has revealed this.
2. I know certainly on the basis of human faith that Tokyo is a city in Japan — because competent and honest geographers have said so.
3. I know as probable, on the basis of human faith, that I will not die within the year of a heart attack — since a competent physician has carefully examined me and assured me that my heart is in a healthy condition.
4. I know with certitude and as immediately self-evident that man is a rational animal — for *rational animal* is the very definition of man.
5. I know certainly and immediately, but not as self-evident, that my

typewriter is gray — for I see it now as immediately present to my sense of sight.
6. I know certainly as mediately true that every triangle has three angles equal to two right angles — for I see that this follows from the necessary truths that every plane figure bounded by three straight lines has three angles equal to two right angles and that every triangle is a plane figure bounded by three straight lines.
7. I know as probable and mediately true the fact of limited evolution — for this is established in the light of very probable premises.

IV. Truth and Falsity

Truth is a conformity between what *is known* and what *is*. Falsity, on the other hand, is a lack of this conformity. On the level of simple apprehension the intellect has not yet committed itself to the way in which things are, nor even to the fact that they are. Hence, there is, in the strict sense, neither truth nor falsity on this level. It is absurd to say "true" or "false" about objects such as *man, green-eyed monster,* and *sleek fast racing horse*. On the level of judgment there is a commitment (either by affirmation or denial) to the existential situation. This commitment can be said to be true or false, according as it does or does not measure up to things as they are. For example, *Every man is able to talk* is true because it expresses things as they really are, and *Every man is a green-eyed monster* is false because it fails to conform to things as they are. In the strict sense then, truth and falsity are found only in the judgment, in the intellect composing or dividing.

V. The Proposition and Its Elements

The object of the second operation is interiorly expressed in a mental proposition. Its verbal expression is an oral or written proposition. A proposition is technically spoken of as a composite expression. But a proposition is a special kind of composite expression. There are others, for example, definitions and arguments.

Composite expressions are basically divided into imperfect and perfect. An imperfect composite expression signifies an incomplete thought which fails to satisfy the intellect. A perfect composite expression signifies a relatively complete thought which does not leave the intellect essentially in suspense. *Rational animal* by itself is an incomplete expression which calls for something more to satisfy the mind. The same is true for a composite expression such as *while the band played on*. These are imperfect composite expressions because they leave the intellect essentially in suspense. On the other hand, there are several types of perfect composite expressions. Deprecative statements (prayers), imperative statements (commands), interrogative statements (questions), and vocative statements (salutations and appeals) are all examples of perfect composite expressions. None of these leaves the mind in suspense as does an imperfect composite expression. Yet none of these is a proposition, for a proposition is that special kind of composite expression which signifies what the intellect knows in judgment. We have seen that the intellect possesses truth or falsity in its second operation depending on its conformity to reality. But it makes no sense to speak of deprecative, imperative, interrogative, and vocative statements as true or false. *Come here!* does not leave the intellect in suspense, but it is neither true nor false. *All men are rational* not only does not leave the intellect in suspense but it can also be said to be true. The proposition can thus be distinguished from other perfect composite expressions by being defined as a composite expression which signifies the true or the false.

To judge is to affirm or deny one thing of something else. That about which something is affirmed or denied is called the subject of the proposition. What is affirmed or denied of the subject is its predicate. In the verbal proposition the connection between the predicate and its subject is most clearly signified by a copula which is some form of the verb "to be." The verb "to be" can be used as a copula distinct from the predicate and the subject. Then both the predicate and the subject are expressed basically as nouns (i.e., *after the fashion* of substances). The subject is *always* basically a noun, but sometimes the predicate is expressed as a verb (i.e., a term which signifies in the mode of existence and action) which signifies in one term both the predicate and the copula. The difference between these two formulations of the proposition is that the former has only a verb-copula while the latter has a verb-predicate. "Some men are great artists" has a verb-copula linking the two nouns which signify respectively the predicate and the subject. "Some men dare to tempt God" has a verb-predicate which simultaneously signifies the predicate and its link with the subject of which it is said. From one point of view the elements into which the proposition is resolved are the noun and the verb. From another point of view they are the subject, predicate, and copula.

For the purposes of rational discourse it is necessary to distinguish subjects and predicates clearly from one another and to recognize each for what it is as a distinct object. Hence it is usually preferable to state an attributive proposition with a verb-copula rather than a verb-predicate. In this way the subject and predicate terms are clearly distinguished from one another and from the copula which unites or separates them in the proposition. In this type of expression it is clear just what the subject-term is and just what the predicate-term is. Propositions stated in this manner are said to be in strictly logical form, and it is always possible to state propositions in logical form even though they are not originally given in this fashion.

Propositions which are not originally in strictly logical form can be put in it by isolating the subject-term and the predicate-term, no matter how complex these may be, and then by joining or separating these through an affirmative or negative form of the verb "to be" in the proper tense. Sometimes the logical subject is grammatically stated in the place of the predicate, and the predicate in the place of the subject. In such cases, of course, it is wise to invert the order of the terms as verbally stated so the grammatical expression is as faithful as possible to the logical expression. The literary quality of propositions in strictly logical form is frequently very poor, and one would not ordinarily speak and write in this mode. However, it is frequently wise for a speaker or writer to recapitulate his assertions and arguments in strictly logical form because they can be most easily logically evaluated in this form. It is always wise for the student beginning his course in logic to recast propositions in strictly logical form. His concern is with logical structure and not literary excellence, and he cannot, especially at the beginning, afford to obscure what is logical in his concern to write well. Consider the following propositions. In each case the proposition is originally stated in a form which is not strictly logical. The second expression of each proposition is in logical form, and it signifies exactly the same thing as the first.

1. **Water boils at 100° centigrade** —
    [Water] is [something which boils at 100° centigrade.]
2. **No elephant ever forgets** —
    [No elephant] is [something which ever forgets.]
3. **Husbands have wives** —
    [Husbands] are [things which have wives.]
4. **Patriotic politicians put the good of their nation above the good of their party** —
    [Patriotric politicians] are [beings who put the good of their nation above the good of their party.]

5. Fools walk in where angels fear to tread —
   [Fools] are [beings who walk in where angels fear to tread.]
6. Pleasant remain the memories of a happy youth —
   [The memories of a happy youth] are [things which remain pleasant.]

Exercise IX

1. Define: judgment, merely existential judgment, attributive judgment, extrinsic evidence, intrinsic evidence, immediate proposition, factually evident proposition, self-evident proposition, mediate proposition, truth, falsity, certitude, probability, noun, verb, subject, predicate, composite expression, perfect composite expression, imperfect composite expression, proposition.
2. What is the difference between simple apprehension and judgment?
3. How can a predicate and a subject with *diverse* definitions be *identified* in an affirmative proposition?
4. *Pinocchio had a long nose.* How can this be understood to be a proposition if the second operation is understood to bear upon existence as well as essence? After all, Pinocchio is only fictional.
5. In what sense is the second operation more excellent even than the third operation of the intellect?
6. Distinguish between the following propositions in terms of motive for assent:
   6.1. The whole must be greater than any one of its parts.
   6.2. The Communists have made great progress in their attempts to conquer the world.
   6.3. Every man is a morally responsible agent.
   6.4. Interplanetary travel will become commonplace by 1985.
   6.5. It rained yesterday.
   6.6. Joe Dimaggio is the greatest baseball player in the history of the game.
   6.7. I am standing up.
   6.8. If I am standing up, I am not not-standing up.
   6.9. No Communist can be trusted to be true to his word.
   6.10. Every organism is a living body.
7. Defend the definition of the proposition as a composite expression which signifies the true or the false.
8. What is the difference between a proposition with a verb-copula and a proposition with a verb-predicate?
9. Wherever possible restate the following as propositions in strictly logical form:
   9.1. John can run the mile in four seconds flat.

9.2. Some men who can be trusted in a tight situation cannot be trusted when things are going smoothly.
9.3. Beware of the dog.
9.4. Happy is the man who owes no debts.
9.5. To the winner go the spoils.
9.6. Don't give up the ship.
9.7. The third world war terrifies all who anticipate it.
9.8. All the world loves a lover.
9.9. No one can say I didn't try.
9.10. Who can fail to admire a courageous man?

CHAPTER X

The Supposition of Terms

I. Signification and Supposition

We have seen that a term is a sign which stands for a quiddity or meaning. The meaning for which the term stands is its comprehension or signification. Since definition expresses the comprehension of concepts or meanings, terms having a common definition have the same signification, and the distinction between univocal, equivocal, and analogous terms is made from this point of view. Univocal terms have identical signification. Equivocal terms have wholly diverse significations. Analogous terms have significations which are basically different but relationally identical. The importance of identity in signification for logical discourse can be seen if we anticipate the third operation for the moment and note the differences in the following attempts at argument:

1. Every triangle has three angles equal to two right angles.
   But, every isosceles triangle is a triangle.
   Therefore, every isosceles triangle has three angles equal to two right angles.
2. Every triangle has three angles equal to two right angles.
   But, John's romantic situation is a triangle.
   Therefore, John's romantic situation has three angles equal to two right angles.

The first argument is obviously good; the second obviously defective. Yet both seem to involve three terms and the same pattern of arrangement. However, in the first argument there is the same signification attached to each term whenever it is used. In the second argument, on the other hand, the same signification is not given to one term, namely "triangle," each time it is used. At best "triangle" in the second argument is only tenuously analogous. Because of this, the argument is formally defective. Its conclusion fails to follow from its premises precisely because the signification of "triangle" varies from premise to premise.

Let us consider a third argument before moving on:

3. Triangle is more universal than isosceles triangle.
But, every isosceles triangle is a triangle.
Therefore, every isosceles triangle is more universal than isosceles triangle.

Something is obviously wrong with this argument. At first glance it might seem that its defect is like that of the second argumentation. Yet examination discloses that the terms have a unity of signification each time they are used. The term "triangle" is defined in the same way each time it is used. However there is *some* way in which "triangle" in the second proposition differs from "triangle" in the first. The term has the same signification in both propositions, but it is used differently in each. In the first proposition "triangle" is used to stand for a concept which is understood as enjoying an existence as an object for the intellect. In the second proposition "triangle" is used to stand for a mathematically real essence which is realized in every individual triangle. The term stands for the same nature each time, but the mode of existence enjoyed by this nature is different in the two cases. In short, the signification is the same, but the term is nevertheless used differently. This difference is sufficient to render this third argument defective. The use which a term has over and above its signification is called its supposition. Terms have signification even on the level of the first operation, but they have supposition only in propositions. Terms in isolation stand for meanings, or else they are not terms. However, terms in isolation cannot be used to stand for one mode of existence rather than another because the first operation of the intellect abstracts from all modes of existence. We define signification as the meaning for which a term stands. We define supposition as the way in which a term is used in a proposition to stand for what it signifies. Although supposition is a property of both subjects and predicates, our concern is primarily with subjects. Predicates do not vary in supposition nearly as much as do subjects. Predicates are usually taken in what we shall call real and personal supposition.

II. Material and Formal Supposition

Supposition is first of all divided into *material* and *formal* supposition. When a term is used simply to stand for itself as a term and not for the quiddity it is designed to signify, it is said to be taken with material supposition. In "Triangle is a three-syllabled word," "Man is easy to pronounce," and "Democracy is semantically loaded," "triangle," "man," and "democracy" all refer to themselves as words. They are used with material supposition.

Whenever a term is used to stand for a quiddity and not for itself, it is said to be taken with formal supposition. Terms in propositions are ordinarily used with formal rather than material supposition. Formal supposition is divided into types, and we will consider its types before examining illustrations of formal supposition.

III. The Types of Formal Supposition

Terms taken with formal supposition are used to stand for quiddities or natures. If the nature for which a term stands is considered precisely as an object existing in the intellect, the term has *logical* supposition. "Man" in "Man is a universal concept" is taken with logical supposition. "Universal concept" can be said of "man" only in so far as this term stands for a concept which has existence in the mind. This is the case for "triangle" in "Triangle is the proximate genus of isosceles figure," for "ability" in "Ability is an abstract concept," and for "animal" in "Animal is more comprehensive and less extensive than organism." On the other hand, if the nature for which a term stands is regarded as realizable in singular existents outside the intellect, that term has *real* supposition. We will subdivide real supposition shortly and take up several examples at that time. Here it is sufficient to note that "man" is taken with real supposition in both "Man is rational" and "Some man is happy."

A third type of formal supposition differs from both logical and real supposition. At the same time, however, it seems to partake of both. Hence we shall call it *logico-real* supposition and shall explain it by reference to the two previous types. As noted, logical supposition results when a term is used to signify a universal nature regarded precisely as an intelligible object existing only in the mind and possessing, as known, a positive unity. For example, in "Man is a species," "species" is said of "man" only in so far as the latter signifies a concept which has positive unity and is regarded exclusively as an object of thought. Here the term "man" has logical supposition. Real supposition, on the other hand, results when a term is used to signify a nature regarded in its capacity as being realizable in singular existents, that is, in subjects which exist outside the mind. For example, in "Man is rational," "rational" is predicated of "man" is so far as it signifies a nature which can exist not only as an object of thought but also in concrete things. Here the term "man" has real supposition. "Man," taken in this sense, is open to individual existents such as Mary and Peter, all of whom can be called "rational."

Now consider the proposition "Man is God's finest earthly creature." Here, as in logical supposition, "man" refers to an object which possesses a positive unity. But at the same time it refers to that object as exercising real, not merely mental existence, for "Man is God's finest earthly creature" is true whether man is known or not. In this it differs from logical supposition, which regards a concept precisely as known. "Man is a species" is true only of the concept man as known by the mind. By referring to concrete existence, then, the term "man" in this proposition bears a likeness to real supposition. However, if a term is taken with real supposition it is open to singular existents, and "man" in "Man is God's finest earthly creature" is taken with a positive unity which precludes this. We cannot say of each man that he is God's finest earthly creature, as we might say of each that he is rational. In this, then, the term "man" here differs from the same term taken with real supposition. But since it does partake of both logical and real supposition in some aspects, we can say that it is here taken with logico-real supposition.

IV. The Types of Real Supposition

A term is taken with real supposition when it signifies a nature or quiddity which is understood as being capable of existence in singular inferiors. But this nature can be regarded in two ways. First, it can be regarded absolutely, that is, in itself, without explicit reference to the individuals in which it is realized. Second, it can be looked at primarily in its reference to those individuals. Thus there are two basic types of real supposition, spoken of as *absolute* and *personal*. Absolute real supposition results when the term is taken primarily for the universal nature itself and only virtually for the individual existents which might participate in that nature. Personal real supposition, on the other hand, occurs when the term is taken primarily for those individual existents, and through them for the universal nature in itself.

There is one important point to notice about personal and absolute supposition. Terms taken with absolute supposition can be used only in propositions with necessary predicates, whereas terms with personal supposition can be used in propositions with either necessary or contingent predicates. "Man" in "Every man is rational," in "Every man is risible," and in "This man is tall" is taken each time with personal supposition.

The difference between "man" in "Man is rational" (where "man" is taken absolutely) and "Every man is rational" (where "man" is taken personally) is that "man" in the first proposition signifies the nature man directly and in the second it directly signifies the individual subjects which possess the nature man.

V. Universal, Particular, and Singular Terms

Terms taken with personal supposition signify natures or quiddities as realized in singular inferiors. However, these terms can signify these natures taken either in their full extension or with some limitation to that extension. Terms used to stand for natures taken with some restriction placed upon their full extension can signify a nature understood with a determinate limitation placed upon their full extension or with an indeterminate limitation. "Man" in "Every man is mortal" is used to the fullest as far as extension is concerned. In this case the predicate is identified with all the inferiors of the subject. When a term is used in this way it is said to be *distributed* or *universal*. "Man" in "Some man is happy" signifies the extension of the nature man as limited indeterminately. Clearly it is not used to the fullest, but just as clearly there is no determinate indication as to the exact limits of the restriction placed upon its extension. This is the *particular* or *undistributed* use of the term. "Man" in "This man is tall" signifies a determinate limit of the full extension of man. Once again there is clearly a restriction placed upon the full extension of the term as used, but here there is a determinate indication of the limits of that restriction. The term "points" to the segment of its extension which is pertinent to the proposition. Terms taken with a determinate limitation placed upon their full extension are said to be *singular*.

We shall discover that we must be able to classify both the subjects and the predicates of propositions as universal, particular, and singular. We shall refer to this as a classification of the extension of these terms. The extension of the subject-term is usually indicated by some modifying word such as "every," "some," or "this." The extension of predicate-terms is not so explicitly identified. However, as we shall see in the next chapter, it is fairly simple to determine the extension of any predicate.

In regard to each of the types of personal supposition, there are a few important points to note. In one sense all terms are universal whether they are used in propositions or not, because they stand for universal concepts. This is true even of those terms which are used particularly and singularly in propositions. We have already defined the universal term as that which can be said of many. When we speak of universal, particular, and singular terms here we are speaking of the use of the universal terms. We can speak of a universal term as *used universally* (in its complete extension) or *particularly* (with an indeterminate limitation to that extension) or *singularly* (with a determinate limitation to that extension). Terms taken with absolute, real supposition do not, strictly speaking, have any extension as used. However, they are virtually universal and can be handled in logical discourse as though they were universal. Whatever belongs intrinsically to a nature or is necessarily consequent upon what belongs intrinsically to a nature must certainly belong to *all* the subjects within the extension of that nature.

Students frequently find it difficult to understand what is meant by the particular use of a term. Perhaps the reason is partially terminological. We say that a term is particular when it is used with an *indeterminate* limitation placed upon its full extension. This is to say that it is used *with no particular segment of its extension in mind,* using the term "particular" in the colloquial sense. Particular subjects are characteristically indicated by the word "some." However, any word can be used so long as it indicates an indeterminate limitation placed upon the full extension of the term. "Certain men," "several horses," and "a few sailors" are all particular terms. The shades of meaning which distinguish these modifiers from one another are not of concern to the logician. Each is on a par with the other as an indication of an indeterminate restriction upon the extension of the term. For the same reason "some triangle" and "some triangles" are logically equivalent; what is logically significant in each case is only the indeterminate limitation of the extension of "triangle." Grammatical singularity and plurality neither add to nor subtract from the status of the term in so far as it is used as particular in logic. For the sake of logical discourse it makes no difference whether a term used particularly refers to one or more subjects. The important point is that it refers to some (i.e., at least one) extensive subject of a given concept without determinately pointing to any definite one.

The singular use of a term in logic should not be confused with the grammatical singular. A term is used singularly in logic if and only if there is a *determinate* limit of the full extension of the term. Thus both "this man" and "these men," and "that tree" and "those trees," have singular supposition. All universal terms modified by demonstratives are taken in singular supposition. This is also true of proper names. Terms taken with logico-real supposition do not, strictly speaking, have any extension. However, they function in logical discourse as though they were singular because they are understood as signifying natures which have a positive unity. Terms with logical supposition can also be handled like singulars for exactly the same reason.

VI. Collective and Divisive Terms

We have divided terms taken with personal supposition into universal, particular, and singular. From another point of view they can be divided into divisive and collective terms. A term is used divisively when it stands for its inferiors taken one by one. It is used collectively when it stands for its inferiors as taken together, i.e., as a group. The term "man" (like all terms) is ordinarily used divisively. This is the case whether it is expressed grammatically in the singular or in the plural. "Every man is subject to death" and "All men are subject to death" are identical propositions. In each case *subject to death* is affirmed of all individual men and of each one of them taken individually. There is no suggestion, even in the second expression, that the predicate is to be understood as a characteristic of all men taken together as one group. In both cases the subject is taken divisively, although this is probably clearer in the first expression than in the second. However, terms which are ordinarily used divisively can be used collectively. When it is said that men may overpopulate the world, or that several horses ate all the hay in the barn, or that the ball players filled the clubhouse, or that our soldiers won the war, "man," "horse," "ball player," and "soldier" are all used collectively. Only as a group can men be thought to overpopulate the world. Only as a group could the horses have eaten all the hay. It is impossible for each of many men to overpopulate the world. And it is impossible for each of several horses to have eaten all the hay. Similarly for "ball player" and "soldier" as used above. Since terms which are ordinarily used divisively can be taken collectively, we must be careful in our use of terms. To confuse the collective and divisive use of terms can be logically disastrous. Whatever can be said of a group of individuals need not be said of each of the individuals in the group. An excellent team may have a player or two who are less than excellent. Whatever can be said universally of the individuals in a group need not be said of the group taken as such. Though all the United States senators, divisively taken, represent one state, the group as a group represents all fifty states.

A term like "man" can be realized in an individual (e.g., Peter or Mary). A term like "army" is limited to a group of individuals (i.e., a number of soldiers taken precisely as a body of men). Although the difference between "man" and "army" might seem like the difference between the divisive and the collective uses of a term, it is a difference in signification, not in supposition. It is by definition that man can be said of individuals, and it is by definition that army can be said only of a group of individuals. A question of definition, of course, is a question of signification and not of supposition. As a matter of fact "army" can be used both divisively and collectively. It is divisive in "Every army needs a good commander in chief." It is collective in "The armies of the nation defended the nation's many fronts."

VII. Possible and Actual Existence

Terms taken with personal supposition can stand for natures understood to exist either actually or potentially. Propositions with universal subjects and necessary predicates may express only possible existence, and ordinarily they are proposed in this fashion. Such propositions are equivalent to propositions with subjects taken with absolute supposition. "Every man is capable of speech," "Every animal is sentient," and "Every triangle is a polygon" are examples of propositions which ordinarily express only possible existence for their subjects. Understood in this way they are true so long as there is a necessary connection between their predicates and their subjects even though no actual instance of their subjects exists. Propositions with particular subjects and contingent predicates may express the actual existence of the subject, and they are frequently intended to do so. "Some trees are tall," "Some men are Americans," and "Some animals are ferocious" are examples of propositions which ordinarily express actual existence. When these propositions are understood to express the actual existence of their subjects they are true only so long as their subjects do actually exist and have the determination which is signified by the predicate. However, propositions with universal subjects can express actual existence. This is ordinarily the case when they are used with contingent predicates, for example, "All men in the second half of the twentieth century live under the threat of nuclear devastation." It can even be the case when they are used with necessary predicates. A proposition with a universal subject and a necessary predicate can suppose the actual existence of the subject. For example, "Every man (and there are men) is capable of speech." On the other hand, propositions with particular subjects and contingent predicates can be understood sometimes to express merely possible existence. "Some trees are tall" can be proposed to mean that there are trees actually in existence and that at least some of these are tall. However, it might also be proposed simply to signify that the nature of trees is such that there is the possibility of their being tall. In the first case the subject-term signifies the nature taken personally and as actually existing. In the second case it signifies the nature taken personally but with no more than possible existence intended. Propositions with a particular subject and a necessary predicate can be enunciated (e.g., "Some man is rational") and these too can express either the actual or the possible existence of their subjects.

VIII. Nonsupposing Subject-Terms

We have noted that both subject-terms and predicate-terms have supposition but that the problem of supposition arises primarily in reference to the subject. The supposition of the subject of a proposition is limited by its predicate and by the copula of the proposition. A predicate may be such that it can be identified with a subject only if that subject is understood to exercise a special kind of existence. For example, the attribute of species can be predicated of a subject only in so far as this subject is understood to exist as a universal object of simple apprehension. We can say a "Man is a species," but we cannot say "John is a species." No singular subject could be legitimately used with this predicate, for no singular subject could stand for something which enjoys the kind of existence relevant to this predicate. The singular subject used with this predicate simply lacks supposition. It can be described as a nonsupposing subject. In a similar fashion any subject which does not meet the demands of the copula of the proposition is a nonsupposing subject. The copula of a proposition which expresses actual existence can be either in the present, past, or future tense. In each case the terms in the proposition must stand for things enjoying the kind of existence demanded by the copula. The subject of "Babe Ruth was a great batter" stands for a man who existed in the past. Thus it meets the demands of this copula, which is in the past tense. The subject of "Babe Ruth is a great batter" or of "Babe Ruth will be a great batter" does not measure up to the requirements of the copula. "Babe Ruth" is a name which can stand only for a figure in the past. It cannot stand for a subject which measures up to demands of a copula in either the present or future tense. When it is expressed with these copulas, it is a nonsupposing subject. Every proposition with a subject which ought to stand in a certain way for something but cannot do so and be true to the demands either of its predicate or its copula has a nonsupposing subject. It is impossible that any affirmative proposition with a nonsupposing subject be true. The predicate of an affirmative proposition is identified with its subject by way of its copula. This identification cannot be realized so long as the subject as used in the proposition is not proportioned to the predicate and the copula. "John is a species" must be false, since "John" cannot stand for a subject which enjoys the kind of existence demanded by the predicate "species." "Napoleon will be a great general" must be false, since "Napoleon" cannot stand for a subject in the future time demanded by the copula in the future tense.

IX. The Unity of Terms and Logical Discourse

At the beginning of this chapter we noted that an argument which is apparently valid can be defective if any of the terms used in it changes its signification or supposition. Now that we have distinguished between the types of supposition let us return to a more detailed consideration of our original point regarding the unity of terms in logical discourse. This discussion can be meaningful to us, although we have not yet come to the third operation of the mind. All that we need to note here concerning logical discourse is that it involves various terms which are used more than once. That *one and the same* term be used more than once is essential to valid argumentation. The terms must stand for the same thing, both from the point of view of signification and supposition, if they are to be said to be identical terms. And if they are not *one and the same*, the attempted discourse is necessarily defective.

1. Signification

Univocal terms stand for precisely the same nature and offer no difficulty in discourse. Equivocal terms stand for absolutely different natures and must be treated as completely diverse terms. Thus equivocation always renders logical discourse defective. Analogous terms are basically diverse in meaning, but they are relationally the same. Because they are not absolutely diverse in meaning they can be used in valid discourse, but not precisely in the way in which univocal terms can be used. We shall, however, omit the discussion here in favor of a consideration later in a metaphysics class. The problem of the use of analogous terms in argumentation is a special problem of method in metaphysics. Metaphysics is a science in which the significant terms are analogous.

2. Material supposition

A term taken with material supposition stands only for itself as a word, and not for the nature ordinarily signified. Hence we cannot, in a valid argument, use the same term once in its material supposition and then in its formal supposition. If a term is taken once with material supposition in a given discourse it must be taken materially each time it appears in this discourse. I cannot argue that Tom (or Harry) is three-lettered because Tom (or Harry) is a man and man is three-lettered.

3. Logical supposition

Whatever is said of a subject expressed by a term taken with logical supposition is said of it *exclusively* in virtue of its existence in the mind as an intelligible object. Thus it is impossible to use a term once with logical supposition in a given discourse and a second time with some other supposition. I cannot argue that man is the definition of man because rational animal is the definition of man and every man is a rational animal.

4. Logico-real supposition

Whatever is said of a subject taken with logico-real supposition is said of it both in virtue of the real situation and in virtue of the positive unity which it has only as an intelligible object existing in the mind. Because of the first point a term cannot be used once with logico-real supposition and a second time with logical supposition in the same discourse. Because of the second point a term cannot be used once with logico-real and then with real supposition. Clearly, then, if a term is used once with logico-real supposition, it must be used in the same way each time it is used in the same discourse. I cannot conclude that Tom is the campus ideal because the football player is the campus ideal and Tom is a football player.

5. Real supposition

Real supposition can be coupled neither with material, logical, nor logico-real supposition. Consequently, a term used with real supposition in a given discourse must be so used every time it appears in that discourse.

6. Absolute and personal supposition

A term used absolutely once may, of course, be used absolutely a second time in any given discourse. And a term used with personal supposition once may be used a second time with personal supposition. However, because terms taken with absolute supposition can be resolved into universal terms taken personally (in propositions expressing possible existence), it is sometimes possible for a term to be used with absolute and with personal supposition in the same discourse. Thus, if I know that man (absolutely taken) is rational, and that John is a man (personally taken), I can legitimately conclude that John is rational.

7. Universal, particular, and singular terms

Discourse can be legitimate even though there is some difference in the way in which one term is used in a given discourse so long as the term is generally taken with personal supposition each time. There is no reason why I cannot conclude that all men are sentient from the fact that all animals are sentient and all men are animals — even though, as we shall see later on, *animal* is used both universally and particularly in this argumentation. We shall take up this point at length in a later chapter.

8. Collective and divisive terms

We have already noted that a term taken collectively is so different from that same term taken divisively that it cannot appear both collectively and divisively in the same discourse. Thus I could not conclude that John numbers over 10,000 because he is a Marquette student and Marquette students number over 10,000.

9. Actual and possible existence

There is no difficulty in this respect so long as the same term is used each time to stand for actual existence or each time to stand for possible existence. There is difficulty only when a term is used once to signify something possibly existing and a second time to signify something actually existing. The general rule which applies in this case is clear. It is legitimate to go from the actuality of something to its possibility, but not from its possibility to its actuality.

Exercise X

1. Define: signification, univocity, equivocity, analogy, supposition, material supposition, formal supposition, logical supposition, real supposition, logico-real supposition, absolute supposition, personal supposition, universal term, particular term, singular term, collective term, divisive term.

2. Distinguish between the signification of a term and its supposition. Why does a term have supposition only in a proposition? Can terms with identical signification have different supposition, and vice versa? Illustrate.

3. What is the reason for our concern with the signification and supposition of terms? What is the importance of these characteristics of terms for logical discourse?

4. Distinguish between material and formal supposition. Illustrate.

5. Distinguish between logical, real, and logico-real supposition. Illustrate.
6. Distinguish between absolute and personal supposition. Illustrate.
7. Distinguish between the universal, particular, and singular use of a term. Illustrate.
8. Discuss some of the terminological difficulties consequent upon the use of the terms "universal," "particular," and "singular" in the context of No. 7.
9. Distinguish between the collective and the divisive use of a term. Illustrate.
10. Distinguish between propositions expressing only possible existence and propositions expressing actual existence. Illustrate.
11. Examine the following pairs of propositions very carefully. Check to see whether or not the term which is apparently used twice in the propositions is used with sufficient unity of signification and supposition to allow for valid discourse. When this is not the case explain the reason why.

    11.1. Army is a collective concept.
    Armies demand expert organization.

    11.2. Armies are different than they used to be.
    An army of ants swarmed by.

    11.3. Army is a concrete term.
    Army is easy to spell.

    11.4. The catcher is the most important man in baseball.
    Eddie is a catcher.

    11.5. Fire is defined as rapid oxidation.
    The salesman who sold me my car was a ball of fire.

    11.6. Every fox has a bushy tail.
    Field Marshal Rommel was a fox.

    11.7. Man is a three-lettered word.
    Tom is a man.

    11.8. Man is God's finest earthly creature.
    Man is a species of animal.

    11.9. Man is a universal concept.
    Every universal concept can be said of many inferiors.

    11.10. Man is subject to death.
    Some men are American.

    11.11. Every square is a polygon.
    This polygon is equilateral.

    11.12. Square is inferior to polygon.
    Every square is equilateral.

    11.13. The outfielders covered the whole outfield well.
    Ted Williams was one of the outfielders.

    11.14. Some dogs are collies.
    The dog is man's best friend.

    11.15. Every one of Santa Claus's helpers is an elf.
    Some of Santa Claus's helpers are elves.

12. Evaluate the pairs of terms in the following argumentations from the point of view of signification and supposition:
    12.1. Since all prisoners hate long terms, and since calisthenics is a long term, all prisoners hate calisthenics.
    12.2. Since organism is superior to animal and man is an organism, it follows that man is superior to animal.
    12.3. Since the Senators are an august body, Senator Wiley is an august body.
    12.4. Since universal is concrete, man is concrete, because man is universal.
    12.5. God is inconsequential, because God is immaterial, and what is immaterial is inconsequential.

CHAPTER XI

The Categorical Proposition

I. The Difference Between the Categorical and Compound Proposition

The proposition is the complex expression which signifies the object known in the second operation of the intellect. In its fundamental form it is composed of a subject-term, a predicate-term, and a copula (the verb "to be") which unites or separates these terms. This fundamental type of proposition has one predicate which is affirmed or denied of one subject. It is called a *simple* or *categorical* proposition. A more complicated type of proposition can be constructed in which the elements are propositions rather than terms and in which the copula is a conjunction rather than the verb "to be." This type of proposition is spoken of as *compound*. *Every animal is sentient* is a simple or categorical proposition. *Every plant is nonsentient* is also a categorical proposition. Each of them is resolved into elements which are terms, with a copula which is the present tense of the verb "to be." *Every animal is sentient, and every plant is nonsentient* is a compound proposition immediately resolved into elements which are themselves categorical propositions and a copula which is not the verb "to be" but the conjunction "and."

Categorical propositions are defined in terms of one subject and one predicate. In the case of *John is happy* there is obviously one predicate and one subject. This is clearly a categorical proposition. The same is true of *Mary is happy*. *John is happy, and Mary is happy* is as obviously a compound proposition. But what of *John and Mary are happy*? If this means *John is happy and Mary is happy*, it is a compound proposition, for *happy* is effectively said twice of two different subjects. But if instead the meaning is *John and Mary are a happy couple*, then the proposition is categorical, for the predicate is said of the two individuals only in so far as they are taken together as one pair. In this case there are two real subjects but only one logical subject, and one predicate. Accordingly this proposition is categorical.

II. Simply Attributive and Modal Propositions

Not all predicates are related to their subjects in the same way. Some are related necessarily to their subjects, others only contingently, some impossibly, others possibly. Some categorical propositions express the connection between subject and predicate without explicitly indicating the mode of this connection (e.g., *Every man is rational*). These are called simply attributive propositions. Others explicitly indicate the way in which the subject and predicate are connected (e.g., *Every man is necessarily rational*). These are spoken of as modal propositions. The expression of the way in which the subject and the predicate are joined in the modal proposition is called the mode. The rest of the modal proposition is called the dictum. A simply attributive proposition is true if it measures up to the factual situation. A modal proposition is true only if both its dictum and its mode are realized. If John is, as a matter of fact, sitting down, the proposition *John is sitting down* is true. However, the proposition *John is necessarily sitting down* is false. The dictum of this modal proposition measures up to the facts, but the mode is false because John might not be sitting down. Propositions which express the mode of necessity are directly denied by those expressing the mode of contingency, and vice versa. Those which express impossibility are directly denied by those expressing possibility, and vice versa. The following examples should help to clarify the differences and relationships between the different types of modal propositions. Some of the examples are true propositions, and some of them, of course, are false.

MODE OF NECESSITY	MODE OF CONTINGENCY
John is necessarily sitting down.	John need not be sitting down.
It is necessary that every man be rational.	A man might not be rational.
Every man must be free.	Not every man need be free.

MODE OF IMPOSSIBILITY	MODE OF POSSIBILITY
No man can be unhappy.	A man might be unhappy.
It is impossible for John to be here.	John could be here.
Angels cannot be weighed.	Angels can be weighed.

III. The Division of the Proposition by Way of Quality

The division of the categorical proposition into the affirmative and the negative is spoken of as a division by way of quality. In an affirmative proposition the predicate is said of its subject. In a negative proposition the predicate is denied of its subject. A proposition is affirmative or negative in quality on the basis of the affirmative or negative character of its copula. In the affirmative proposition the copula unites the predicate with the subject (e.g., *Some man is happy*). In the negative proposition it separates the predicate from the subject (e.g., *Some man is not happy*). Subject-terms and predicate-terms can themselves be either affirmatively or negatively expressed, but this does not determine the quality of the proposition. The following are all affirmative propositions:

> Every man is rational.
> Every brute is nonrational.
> Some unhappy men are unmarried.
> Some baseball players who are not excellent hitters are of more value to their team than those who are excellent hitters but cannot field well.
> Every A is B.
> Some non-A is B.
> Every non-A is non-B.
> Some A is non-B.

The following, on the other hand, are all negative propositions:

> No horses are nonsentient.
> Many Americans are not rich.
> Some non-Catholics do not fear a Catholic president.
> Some of those below deck when the blast occurred were not in the group unable to save themselves.
> No A is B.
> Some A is not non-B.
> Some non-A is not B.
> No non-A is B.

IV. The Division of the Proposition by Way of Quantity

Categorical propositions are divided into the universal, the particular, and the singular on the basis of the extension of their subjects. This is a division of the proposition according to quantity. Propositions with fully distributed or universal subjects are said to be universal in quantity.

Those with undistributed or particular subjects are particular in quantity. Those with singular subjects are singular in quantity. A universal proposition either affirms or denies its predicate of its subject with this subject taken in all its extension (e.g., *Every man is rational, All triangles have angles whose sum is equal to 180 degrees,* and *No organism is immaterial*). A particular proposition affirms or denies its predicate of an indeterminate segment of the extension of the subject (e.g., *Some triangle is isosceles, Not all men are American, Several soldiers marched by,* and *Many terms are difficult to define*). A singular proposition affirms or denies its predicate of a determinate segment of the extension of the subject (e.g., *This pen is mine, Those statements were false, Mary's mother is ill,* and *John is a student*).

One or two additional words remain to be said on the quantity of the proposition. First of all, the quantity of a proposition depends upon the extension of its subject. What is to be done with a proposition whose subject is indefinitely expressed in reference to extension? In such a case it is necessary somehow to determine the extension of the subject before the proposition can be quantitatively classified. In the proposition *Triangles are isosceles* it is clear from the nature of triangle that the subject of the proposition is to be understood neither universally nor singularly but particularly. In the case of *Triangles are three-sided,* however, the subject should be considered as universal. Hence, *Triangles are isosceles* is particular in quantity, and *Triangles are three-sided* is universal. Second, the terms "particular" and "singular" could be misleading. We must recall that technically "particular" refers to an *indeterminate* limitation placed upon the full extension of a concept while "singular" refers to a *determinate* limitation placed upon its full extension. The particular (technically speaking) is used when (colloquially speaking) no one in particular is referred to. And the singular (speaking with logical accuracy) might well be plural rather than singular (speaking in the terms of the grammarian). We have pointed out before that the difference between the grammatical singular and the grammatical plural is of little significance in logic. This is especially true in the case of the particular proposition. From the point of view of logical significance it makes little difference whether we say "Some man is happy" or "Some men are happy." Logic is primarily ordered to science, and what is of scientific importance in each statement is the indication that human nature is open to happiness. In a similar fashion the fact that human nature is open to the absence of happiness is expressed equally well in both "Some man is not happy" and "Some men are not happy." Finally, in the strict sense, only terms taken with personal supposition are divided into the universal, particular, and singular. Thus, propositions which have subjects taken with supposition other than personal will not be classified strictly as to quantity. They will, however, have a logical force similar to either universal or singular propositions. Propositions with subjects taken with logical or logico-real supposition function like singular propositions, while those with subjects taken absolutely are virtually universal.

V. The Extension of the Predicate

The extension of the subject of a proposition is ordinarily indicated by some auxiliary term such as "every," "some," or "this." However, this is not the case with the predicate of a proposition. Predicates are never introduced by auxiliary terms indicating the extent to which they are to be taken in the proposition. Yet, it is necessary to know the extension both of the subject and of the predicate of any proposition. As a matter of fact it is as easy to determine the extension of the predicate as it is to determine that of the subject. First of all, with rare exceptions, predicates are not singular. Propositions which apparently have singular predicates are ordinarily not stated in perfect logical form. These should be restated so that the apparent predicate and subject are interchanged. For example, *One of the most important nations in the free world is the United States* in perfect logical form would be stated *The United States is one of the most important nations in the free world*. Predicates, then, are either particular or universal. They are particular in propositions affirmative in quality and universal in propositions negative in quality. The reasons for this are clear, if we consider the nature of affirmative and negative propositions. The affirmative proposition signifies the identification of a predicate with a subject. This identification is primarily from the point of view of comprehension but correlatively, of course, from the point of view of extension. To say *Every A is B* is to assert that the total meaning of B is realized in a subject which is A. That A is comprehensively more than B is possible. But if A can be comprehensively richer than B, then its extension may be less than that of B, for we have seen that concepts richer in comprehension have less extension. Thus, from the formal point of view A must be considered less extended than its predicate B and consequently, from the point of view of extension, it is identified with only some of B. Thus the predicate of an affirmative proposition is fully identified with its subject from the point of view of comprehension, only partially as far as extension is concerned. It must, therefore, be classified as a particular term. This is clearly the case with *Every man is animal*. Since brutes as well as men are animal, *man* cannot exhaust the extension of *animal* but can at best be considered as identified with some of it. This is also the case with *Every triangle is a plane figure* and *Every man is biped*. But what of *Every man is a rational animal?* As a matter of fact the predicate and subject in this proposition are coextensive so that *man* does exhaust the extension of *rational animal*. This is true because of the intelligible content of the predicate, but not because of the formal structure of the affirmative proposition. The definition of a species, the specific difference of a species, and a specific property of a species are, as a matter of fact, coextensive with the species. Whenever these are predicated of the species the predicate happens to be used fully as far as extension is concerned. However, this is something accidental to the nature of the affirmative proposition. The formal structure of the affirmative proposition *allows* for a coextensive predicate but at best guarantees only a particular predicate. It remains true, from the point of view of formal logic, that the predicate of an affirmative proposition must be classified as particular.

The negative proposition expresses the separation of a predicate from a subject. From the point of view of comprehension the negative proposition signifies that the meaning of the predicate is not fully realized in the subject. Since the meaning of the predicate is the sum of its comprehensive notes, this means that the negative proposition is an assertion that not all the comprehensive notes of a predicate are found in a subject. But nothing falls within the extension of a concept unless it possesses all the comprehensive notes of that concept. Thus the negative proposition separates its subject completely from the extension of its predicate. To say *No A is B* is to assert that not all the comprehension of *B* is found in any *A* and that no *A* is found within the extension of *B*. There are no exceptions possible here even on the part of material or factual condition. Without exception the predicate of a negative proposition is universal.

A proposition can be so expressed that it stresses either the comprehensive or extensive relationship of predicate to subject. Since comprehension and extension are correlative it makes little difference which is stressed so long as there is no intention to suppress the other. To say "Every man is animal" is to stress the fact that the comprehensive notes of *animal* are found in every man. To say "Every man is an animal" is to stress the inclusion of all men within the extension of *animal*. Both express the same truth, and either is acceptable. The same is true for "No horse is man" and "No horse is a man." The first points to the predicate comprehensively taken and the second to the predicate extensively taken. Yet each signifies the same truth, and either is an acceptable expression.

VI. A, E, I, and O Propositions

We divided propositions from the point of view of quantity into universal, particular, and singular, and from the point of view of quality into the affirmative and the negative. Combining these classifications we find six types of propositions; universal affirmative, universal negative, particular affirmative, particular negative, singular affirmative, and singular negative. In order easily to identify these in the future we can assign symbols to stand for the various types. Traditionally logicians have used the letters A, E, I, and O to stand for the different propositions. A stands for the universal affirmative, and I for the particular affirmative (from the Latin word AffIrmo). E stands for the universal negative, and O for the particular negative (from nEgO). Because singular propositions, as we shall see, are frequently used with a logical force similar to that of the universal proposition, but not always and in every way, singular affirmative propositions can be spoken of as A^1 and singular negatives as E^1. The following propositions illustrate the several types:

A — Every triangle is a plane figure.
All who do not love their country are unworthy of its protection.
Every non-A is B.

E — No good Christians are Communists.
None of the moral virtues are possessed by brutes.
No non-A is non-B.

I — Some men are American.
Several human faculties are cognitive.
Certain men are mentally disturbed.
Some A is non-B.

O — Some men are not mathematicians.
Not every Communist is Russian.
All that glitters is not gold.
Some A is not B.

A^1 — This nation is a democracy.
These triangles are equilateral.
John is president of his class.
This A is non-B.

E^1 — This cow is not healthy.
Those men are not students.
John is not wearing a hat.
That A is not B.

THE CATEGORICAL PROPOSITION 127

These examples do not exhaust the possible ways of expressing the various types of categorical propositions grammatically. What is of importance for the logic student is the logical fact of universal affirmation, of universal negation, etc. These may be expressed in many ways grammatically, and it is not the business of a logic text to list all the grammatical possibilities. However, we might note that the universal negative proposition is never rendered grammatically by the explicit use of "every" with the subject and "not" with the copula. "Not" with "every" not only diminishes the force of the copula by making for negative quality, but it also diminishes the extension of the subject and renders it particular. *Not every A is B* and *Every A is not B* are equivalent to *Some A is not B*. The universal negative expression is *No A is B*.

Exercise XI

1. Define: categorical (simple) proposition, simply attributive proposition, modal proposition, mode, dictum, affirmative proposition, negative proposition, universal proposition, particular proposition, singular proposition.
2. Contrast categorical and compound propositions in terms of their respective elements.
3. Distinguish between the dictum and the mode of a modal proposition and explain what is demanded of each for the truth of the modal proposition.
4. Supply original examples for each type of modal proposition. Then directly deny each with the appropriate modal proposition.
5. Explain how a proposition with either a negative subject or negative predicate can be an affirmative proposition.
6. How do universal propositions differ from particular and singular propositions, and how do these latter differ from one another?
7. What does it mean to say predicates of affirmative propositions are used fully in reference to comprehension and partially in reference to extension? Explain why this is so in terms of the formal structure of the affirmative proposition.
8. What does it mean to say predicates of negative propositions are used partially in reference to comprehension and fully in reference to extension? Explain why this is so in terms of the formal structure of the negative proposition.
9. What is the extension of the subject and of the predicate of each A proposition? Of each E proposition? Of each I proposition? Of each O proposition? Of each A^1 proposition? Of each E^1 proposition?
10. Indicate the extension of the subject in the parentheses following the subject, the extension of the predicate in the parentheses following the

predicate, and the classification of the proposition as a whole in the parentheses following the proposition. Use the symbols U, P, S, A, E, I, O, A¹, and E¹.

10.1. No A ( ) is non-B ( ). ( )
10.2. Some triangles ( ) are isosceles ( ). ( )
10.3. No four-legged animals ( ) are human ( ). ( )
10.4. Not every nonessential predicable ( ) is an accident ( ). ( )
10.5. No athletes who are under six feet tall ( ) are well suited to the position of center on the modern basketball team ( ). ( )
10.6. Some men who are not happy in their work ( ) are unable to do it well ( ). ( )
10.7. No disloyal Americans ( ) are worthy of their privilege to vote ( ). ( )
10.8. These numbers ( ) are divisible by two ( ). ( )
10.9. All that glitters ( ) is not gold ( ). ( )
10.10. The Civil War ( ) was fought in the nineteenth century ( ). ( )
10.11. Flies ( ) fly ( ). ( )
10.12. Every non-A ( ) is not B ( ). ( )
10.13. Some A ( ) is not non-B ( ). ( )
10.14. This A ( ) is non-B ( ). ( )
10.15. A ( ) is B ( ). ( )
10.16. Not any A ( ) is B ( ). ( )

11. Suggest examples of propositions with subjects taken with other than personal supposition and discuss the "quantity" of each.

CHAPTER XII

The Compound Proposition

I. The Nature of the Compound Proposition

A compound proposition is a single proposition composed of several component propositions joined by a copula which is a conjunction rather than a verb. The elements into which a categorical proposition are immediately resolved are terms. Those into which a compound proposition is immediately resolved are themselves propositions. In the least complex compound propositions the elements are categorical propositions. In more complex compound propositions the elements might themselves be compound propositions. Thus *The Communist world is militarily strong* is a categorical proposition immediately resolved into the subject-term *Communist world* and predicate-term *militarily strong*. *The Communist world is militarily strong, and the free world is relatively weak* is a compound proposition immediately resolved into the two categorical propositions, *The Communist world is militarily strong* and *The free world is relatively weak*. *If the Communist world is militarily strong, and the free world is relatively weak, then the free world is in danger of annihilation* is a more complex compound proposition immediately resolved into a compound proposition and a categorical proposition. The compound component is *The Communist world is militarily strong and the free world is relatively weak*, while the categorical component is *The free world is in danger of annihilation*. Note that the copula in the categorical example is the present tense of the verb "to be." In the two compound examples the copula is respectively the conjunction "and" and the conjunction "if . . . then."

II. Conjunctive and Hypothetical Propositions

The elements of the compound proposition are propositions, ultimately categorical propositions. Hence, the elements of the compound proposition are logically divided after the fashion of the divisions of the categorical proposition treated in the preceding chapter. Compound propositions themselves, however, have divisions proper to themselves. First of all, some compound propositions express a definite commitment to the truth-status of each of their elements. The proposition *The Communist world is militarily strong, and the free world is relatively weak* asserts the truth determinately of both of its categorical components. On the other hand, the proposition *If the Communist world is militarily strong, then the free world is in danger* does not make a definite commitment to the truth of either of its elements. It does, however, involve a commitment to the nexus or link between these component parts. This commitment requires one to admit that if the first is true the other must also be true. The link or nexus between these two parts does not depend upon their truth, but it does exclude the possibility of the first element being true while the second is false. Those compound propositions which involve a definite commitment to the truth-status of each element are conjunctive propositions, whereas those which do not involve this commitment are hypothetical propositions.[1]

III. The Copulative Proposition

The most elementary type of conjunctive proposition is the copulative proposition. The copulative proposition simply asserts the truth of several propositions joined by the conjunction "and" or its equivalent. The copulative proposition expresses nothing more than the simultaneous truth of several co-ordinate propositions. As such, a copulative proposition is true when each of its elements is true. It is false if any one of its elements is false, regardless of the truth value of the others.

Propositions which are so constructed that they clearly involve several co-ordinate propositions joined by "and" or its equivalent are said to be openly copulative. This is the case with *Every triangle has three sides, and some triangles have equal sides.* Propositions which do not clearly and immediately reveal their commitment to the truth of several co-ordinate propositions are said to be occultly copulative. Because occultly copulative propositions can be expanded and expressed as openly compound they are called exponible propositions. *Only the good die young* is an occultly copulative proposition which can be expanded to *The good die young, and none who are not good die young.* An exponible

---

[1] Many logicians call all compound propositions hypothetical. Others call only conditional propositions hypothetical. Our use of the term "hypothetical" is more restricted than the former and less restricted than the latter. Our use of the term "conjunctive" is also different from the sense given it in many logic texts, where it is used to signify the type of proposition called "disjunctive" in this book. The term "alternative" is used here to refer to propositions spoken of in most other texts as disjunctive.

proposition is true, of course, only if each of the propositions which appear in its expanded form is true. It is false if any one of them is false, regardless of the truth of the others.

The preceding example of an occultly copulative proposition is called an exclusive proposition because it identifies its predicate with its subject to the exclusion of any other subject. A second type of occultly copulative proposition joins a predicate to a subject which is understood to involve exceptions. This is the exceptive proposition. A third type of occultly copulative proposition is called the reduplicative proposition. The reduplicative proposition expresses the precise aspect of the subject in virtue of which the predicate is either affirmed or denied of it. The following are examples of exponible propositions, together with the openly copulative propositions to which they might be expanded and in terms of which they must be understood for the purposes of verification:

EXCLUSIVE

God alone has the power to create.

God has the power to create, and nothing other than God has the power to create.

Only lovers can adequately define love.

Lovers can adequately define love, and no one who is not a lover can adequately define love.

EXCEPTIVE

Everyone except Truman expected Dewey to win the presidency in 1948.

All men other than Truman expected Dewey to win the presidency in 1948, and Truman did not expect Dewey to win the presidency in 1948.

None of the American League teams with the exception of the Yankees has been a consistent pennant winner.

No American League team other than the Yankees has been a consistent pennant winner, and the Yankees are an American League team, and the Yankees have been a consistent pennant winner.

REDUPLICATIVE

Every priest as a man suffers temptation.

Every priest is a man, and all men suffer temptation, and every priest suffers temptation precisely in virtue of being a man.

As a citizen you are entitled to vote.

You are a citizen, and every citizen is entitled to vote, and you are entitled to vote precisely as a citizen.

IV. Adversative and Causal Propositions

Copulative propositions assert the simultaneous truth of several co-ordinate components. Other conjunctive propositions express not only the simultaneous truth of their component propositions but the subordination of one component proposition to another. In an adversative proposition the subordinate component is shown to be somehow opposed to the main component. This opposition is verbally expressed by words such as "although," "despite," and "whereas." *Truman was elected in 1948, despite the fact that no pollster had predicted his victory* is an example of an adversative proposition. It is true only if Truman was elected in 1948 and if no pollster had predicted his victory *and* if there is some suggestion of opposition between the latter truth and the former. Other examples of adversative propositions include:

> Although contraries seem like pure and simple opposites, they might be false together.
>
> The weather turned out to be beautiful, despite an adverse weather report in the morning.
>
> John won the race, even though he lost his shoe twenty yards from the finish.

A second type of conjunctive proposition which expresses the subordination of one element to another is the causal proposition. In the causal proposition the subordinate component gives a reason for the truth expressed in the main component. The subordination of the causal component to the main component is expressed by words such as "because" and "since." *Human beings are morally responsible agents, because they are possessed of free will* is a causal proposition. It is true only if each of its elements is true *and* if the subordinate element is truly related to the other after the fashion of cause to effect. Additional examples of causal propositions are:

> Logical relations are not able to be categorized, because they are beings-of-the-reason.
>
> This is a compound proposition, for it has a causal clause added to a main clause.
>
> No one understood the visitor, since he spoke only in French.

Adversative and causal propositions are conjunctive. Like copulative propositions they assert determinately the truth of each of their elements. But unlike copulative propositions they also assert the subordination of one element to another. Thus more is required to verify adversative and causal propositions than copulative. For the copulative proposition the truth of each element suffices. For the adversative and causal propositions the truth of each component *plus* the realization of the adversative or the causal nexus is necessary. Thus, adversative and causal propositions can be thought of as adding something to simply copulative propositions. Each might be generally described as a copulative proposition *plus*.

V. The Conditional Proposition

As noted above, hypothetical propositions, unlike conjunctive propositions, do not assert determinately the truth of each of their elements but only express a commitment to a nexus or link between them. The example we used of a hypothetical proposition was *If the Communist world is militarily strong, the free world is in danger.* This is a conditional proposition. The conditional is not the only kind of hypothetical proposition. However, it is the most fundamental, and we shall treat of it before the others. A simple conditional proposition is an assertion that two propositions are so related that the truth of one of them (the antecedent) necessitates the truth of the other (the consequent). The commitment of the mind in the simple conditional proposition is neither to the truth of the antecedent nor to the truth of the consequent. The simple conditional commitment bears only on the nexus between the components, namely, that the antecedent cannot be true without the consequent being true as well. Thus the verification of the simple conditional does not depend upon the truth of its elements. *If the Communist world is militarily strong, the free world is in danger* is true not because the Communist world *is* militarily strong (or is *not* militarily strong) nor because the free world *is* in danger (or is *not* in danger). This proposition is true because its antecedent and consequent are so related that if the antecedent were true the consequent would necessarily be true. The following examples illustrate the simple conditional:

> If Marxist philosophy is true, the free world is doomed. (This is verified because Marxist philosophy teaches the inevitable universal success of Communism.)
> If contingent beings exist, a necessary being exists. (This is verified because only a necessary being can explain the existence of any being whatsoever.)
> In baseball, if a right hander is pitching, the odds favor a left-handed batter. (This is verified because a ball thrown by a right-handed pitcher curves in to a left-handed batter and is thus better able to be seen by him than by a right-handed batter.)

> If it is eleven o'clock, the mail has been delivered. (This is verified as long as the habitual practice of the mailman is such that he never delivers the mail later than eleven o'clock.)

These examples illustrate two things. First of all, they show that the commitment involved in a conditional proposition bears essentially upon the connection between antecedent and consequent. Second, they illustrate a variety of ways in which this connection can be guaranteed by the reality of things. The verification of the simple conditional depends not on the truth of its elements but entirely on the implicative nexus between them. This nexus exists as long as the component propositions are so related that the antecedent cannot be true without involving the truth of the consequent as well. The scientific value of the conditional proposition depends on the degree of necessity in the foundation for the nexus.

Thus far we have spoken only of simple conditional propositions, which assert that an antecedent cannot be true without a consequent being true. There is a stricter type of conditional proposition, the reciprocal conditional proposition. The latter asserts as much as the former and, in addition, affirms that the consequent cannot be true without the antecedent being true as well. In a simple conditional proposition we assert that a given proposition implies a second proposition, meaning that the first cannot be true without the second being true also. In a reciprocal conditional we assert that a given proposition and only this implies a second proposition. This means not only that the first cannot be true without the second being true but also that the second cannot be true without the first being true as well. Here too the truth of the proposition as a whole does not depend upon the truth of its elements but solely on the nexus of implication between them. The difference between the simple and reciprocal conditional propositions is simply that in the latter there is a stricter nexus or link between antecedent and consequent. A simple conditional is true as long as the truth of its antecedent implies the truth of its consequent, although its consequent can be true without involving the truth of its antecedent. A reciprocal conditional is not true under these circumstances. It is true only if both elements reciprocally imply one another. The following examples are illustrations of true reciprocal conditional propositions:

> If and only if man can know the intelligible, the human soul is incorporeal.
>
> If and only if God exists as a creator, contingent beings exist.
>
> If and only if Oregon is north of California, San Francisco is south of Portland.

> If and only if each player on a team does his job well, the team functions perfectly.

Clearly the existential situation which would guarantee the truth of a reciprocal conditional would verify a similar simple conditional, but not vice versa. Hence, if I can say *If and only if a man is virtuous, he is truly happy*, I can surely propose *If a man is virtuous, he is truly happy*. However, I cannot assert *If and only if a man runs, he moves* just because *If a man runs, he moves* is true. The reciprocal commitment includes the simple commitment and more. Hence, we can move from the reciprocal to the simple, but not necessarily from the simple to the reciprocal.

VI. The Alternative Proposition

The conditional proposition is only one kind of hypothetical proposition. A second type is the alternative proposition. The object of the commitment proper to the alternative proposition is a nexus or link between component parts which requires that at least one of these components be true. By this we do not mean simply that the alternative proposition asserts that one of its elements is, perhaps quite independently of the others, necessarily true. Rather we mean that the several elements are so related that they cannot all be false together, so that one, at least, must be true.

The copula in the alternative proposition is signified by the conjunction "or." Generally speaking, this "or" indicates alternate propositions one of which, in the light of the others, must be true. This is all that the copula signifies in the less strict type of alternative proposition, called the inclusive alternative proposition, so called because it includes the possibility that more than one of the elements is true. The elements are alternates only in the sense that, given the falsity of the others, one must be true. The proposition is true only if its elements are so related that, in the light of this relationship, at least one must be true. Here, as with the conditional, the truth of the alternative proposition does not depend on the truth of its elements taken in themselves but upon the truth of the nexus between them. However in this case the nexus is not one directly of implication, but rather of alternation. The following examples illustrate true inclusive alternative propositions:

> Either juvenile delinquency will be curbed or our streets will be safe for no one.
> Either the free world prepares to defend herself or the Communist world will conquer her.

> A generic predicable is either a proximate genus, a remote genus, or an ultimate genus.
>
> **Either a necessary being exists or no contingent beings exist.**

These propositions are true inclusive alternative propositions not simply because one or more of the elements in each happens to be true but because at least one must be true in the light of the others. These propositions illustrate several points concerning alternative propositions. First, they show that alternatives can involve two or more elements. We shall be concerned chiefly with two-membered hypotheticals, but we should note that alternative propositions admit of any number of elements and are logically manageable no matter how complex they may be. Second, like the examples used to illustrate conditional propositions, these examples show that there is a variety of ways in which the nexus between the elements of the proposition is realized. Here we must also remember that these are examples of *inclusive* alternative propositions. This means that they assert merely that one element at least must be true; the others need not be false but may also be true. If this is not crystal-clear from the context then the propositions ought to signify the situation explicitly. Expressed most clearly, then, these examples would be rendered: *Either juvenile delinquency will be curbed or our streets will be safe for no one — or perhaps both, etc.*

The more strict form of alternative proposition is the exclusive alternative. Like the inclusive alternative, the exclusive alternative asserts that at least one of several alternate propositions must be true. Unlike the inclusive, however, it excludes the possibility that more than one of the alternatives can be true. It positively asserts that only one can be true. The elements are alternates in the strictest sense, for given the falsity of the others one must be true, and given the truth of any one the others must be false. Once again, of course, the truth of the proposition as a whole does not depend upon the truth value of its elements taken in themselves, but rather upon the nexus between them. This demands that one and only one be true regardless of which one that is. The following examples illustrate exclusive alternative propositions:

> Either some being is finitely real or no being is able to be categorized.
>
> Either all men are perfect or some men are not what they should be.
>
> Either no men are mentally unbalanced, or all men are mentally unbalanced, or some men are mentally unbalanced while others are not.

Either the South won the Civil War, or the North won the Civil War, or neither side won the Civil War.

These examples illustrate the same points illustrated by the examples offered for the inclusive alternative. But these examples also show that only one of the alternatives can be true for exclusive alternative propositions. Once again, if the context is not sufficiently clear to indicate they are to be taken as exclusive alternatives and not simply as inclusives, this must be pointed out explicitly. Then the propositions would be given as follows: *Either some being is finitely real or no being is able to be categorized — but not both*, etc.

VII. The Disjunctive Proposition

The disjunctive proposition is the last of the hypotheticals for us to consider. The disjunctive proposition asserts that its component propositions are so related that at least one of them must be false. The elements of the disjunctive proposition are propositions whose mutual relationship is signified by a copula which combines the negation "not" with the conjunction "and." The disjunctive proposition thus appears to be merely a denial of an original copulative proposition. This, however, is misleading, for the negation of a copulative proposition asserts no more than the falsity of at least one of the components of the original copulative. The disjunctive commitment, on the other hand, bears directly upon the nexus between its component propositions and asserts that these components are so related that in virtue of this relationship at least one of them must be false. Though disjunctives with more than two elements are possible, ordinarily disjunctive propositions are of the two-membered variety. The following examples illustrate the disjunctive proposition:

> It can not be true that all predicates are superiors and that some predicates are singular.
> Not all Americans are loyal so long as some are Communist.
> A predicate and its contradictory cannot both be affirmed of the same subject.
> It is impossible for a community to be well regulated and its leaders unintelligent.

VIII. Reduction of Alternative and Disjunctive Propositions to the Conditional

The conditional proposition is the most fundamental type of hypothetical proposition. Any alternative or disjunctive proposition can be reduced to a corresponding conditional equivalent in meaning and basic to it. The hallmark of the hypothetical proposition in general is the logical nexus between its component propositions, in virtue of which the truth value of one has a bearing on the truth values of the others. Every hypothetical proposition, then, involves, at least implicitly, the relation of antecedent to consequent. This relation is explicitly and directly expressed in the conditional proposition, which is by definition a direct commitment to a sequential connection between antecedent and consequent. This relation is implied in both alternative and disjunctive propositions. Thus all three — conditionals, alternatives, and the disjunctives — are hypothetical propositions and, as we shall see, all can be used as premises in hypothetical argumentation. Yet conditionals remain the basic type of hypothetical to which the others can be reduced.

The inclusive alternative and the disjunctive proposition are reduced to simple conditionals. The stricter alternative is reduced to the stricter conditional. Thus the exclusive alternative is reduced to the reciprocal conditional. The following examples illustrate the reduction of alternative and disjunctive propositions to equivalent conditional propositions:

INCLUSIVE ALTERNATIVE

Either juvenile delinquency will be curbed or our streets will be safe for no one.

Either the free world prepares to defend herself or the Communist world will conquer her.

SIMPLE CONDITIONAL

If juvenile delinquency is not curbed, our streets will be safe for no one.

If the free world does not prepare to defend herself, the Communist world will conquer her.

EXCLUSIVE ALTERNATIVE

Either some being is finitely real or no being is able to be categorized.

Either all men are perfect or some men are not what they should be.

RECIPROCAL CONDITIONAL

If and only if some being is finitely real, some being is able to be categorized.

If and only if all men are perfect, all men are what they should be.

DISJUNCTIVE

It can not be true that all predicates are superiors and that some predicates are singular.

Not all Americans are loyal while some are Communist.

SIMPLE CONDITIONAL

If all predicates are superiors, no predicates are singular.

If all Americans are loyal, no Americans are Communist.

IX. Categorical Propositions Expressed as Hypothetical

"If anything is a man, that thing is rational" apparently signifies a simple conditional assertion. The same thing is true for "If anything is a plant, it is not sentient." However, these expressions can be taken as equivalent to "Every man is rational" and "No plant is sentient," so long as each of these categorical propositions is understood to intend only possible existence. It makes little practical difference, at least on the level of the second operation of the intellect, whether the expressions are taken as signifying the hypothetical or the universal categorical proposition, since each reduces in meaning to the other. However, there is a difficulty which arises on the level of the third operation if these propositions are understood as hypothetical rather than categorical, since the rules for a valid hypothetical syllogism do not apply without modification to a syllogism based on one of these propositions taken as hypothetical. On the other hand, if these propositions are understood as categorical they offer no difficulties as premises in a valid categorical syllogism. Hence it seems preferable to take these propositions as categorical rather than hypothetical, especially when they function as premises in argumentation.

What we have said here of some apparently simple conditional expressions applies as well to certain inclusive alternative and disjunctive expressions. "Every man is rational" is equivalent not only to "If anything is a man, that thing is rational," but also to "Either something is not a man or it is rational" and "Nothing is both a man and not rational." And "No plant is sentient" is equivalent not only to "If anything is a plant it is not sentient," but also to "Either a thing is not a plant or it is not sentient" and "Nothing is both a plant and sentient." The apparently reciprocal conditional expression "If and only if a being is free, it is a moral agent" and the exclusive alternative expression "Either a being is free or it is not a moral agent" are both equivalent to a copulative proposition composed of two universal categoricals with inverted extremes, namely, "Every free being is a moral agent, and every moral agent is a free being."

X. The Symbolic Representation of Compound Propositions

The constructs of formal logic, especially compound propositions, lend themselves readily to symbolic expression. As long as they represent the logical situation without distortion, symbols can frequently facilitate the discussion of logical procedures. At this time we shall introduce the use of symbols for compound propositions. However, in order not to complicate things too much at this point, we shall limit ourselves somewhat by symbolizing only the openly copulative proposition and the several hypothetical propositions we have considered, namely, the simple conditional, the reciprocal conditional, the inclusive alternative, the exclusive alternative, and the disjunctive proposition. We shall also limit ourselves for the most part to a consideration of two-membered compound propositions, though we have seen that multimembered copulatives, alternatives, and even disjunctives are possible.

We have already used capital letters to stand for the subjects and predicates of categorical propositions (e.g., Every A is B). We shall use small letters now to stand for the elements of compound propositions. Here, of course, the letters will stand for propositions and not for terms, for the component parts of compound propositions are themselves propositions. We shall use a slightly elongated dash (—) as the sign of negation or contradiction, a dot (·) for the conjunction "and," a wedge ($\vee$) to stand for the copula in the inclusive alternative, an inverted wedge ($\wedge$) for the copula of the exclusive alternative, and an arrow ($\rightarrow$) to signify simple conditional sequence, with a two-headed arrow ($\leftrightarrow$) to signify reciprocal conditional sequence. Where punctuation is necessary, especially to insure the correct application of the sign of negation, parentheses and brackets will be used. Finally, an identity in meaning between two propositions will be indicated by the sign of equivalence ($\equiv$). Thus, the copulative proposition *The Braves won the National League pennant in 1958, and the Yankees won the American League pennant that same year* can be symbolized simply as $a \cdot b$. The simple conditional proposition *If contingent beings exist, a necessary being exists* becomes symbolically $a \rightarrow b$. The reciprocal conditional *If and only if a man can know the intelligible, the human soul is incorporeal* becomes $a \leftrightarrow b$. *Either juvenile delinquency will be curbed or our streets will be safe for no one* can be expressed $a \vee b$ because it is an inclusive alternative, while the exclusive alternative *Either some being is finitely real or no being is able to be categorized* is symbolically represented $a \wedge b$. This leaves only the disjunctive to be symbolized. Here, however, there is some difficulty. Were the disjunctive simply the negation of an original copulative we could symbolize it $-(a \cdot b)$. But this, as we have seen, is not the case. However, we can symbolize the disjunctive by modifying the symbolic expression of the negation of the copulative. We do this by adding an "n" to signify the kind of necessity involved in a logically disjunctive proposition. $-(a \cdot b)$ is a copulative proposition which simply denies the copulative $a \cdot b$. But $n-(a \cdot b)$ is a disjunctive proposition which denies $a \cdot b$ precisely because it asserts the impossibility of $a \cdot b$.

XI. The Truth-Functional Proposition and Its Truth Table

A compound proposition is said to be truth-functional when its truth as a whole depends solely upon the truth values of its component parts. This is the case with the copulative proposition $a \cdot b$ and with its contradiction $-(a \cdot b)$. The various possible combinations of truth and falsity for the elements of a truth-functional proposition can be tabulated along with the corresponding truth value of the proposition itself in a schematic arrangement called a truth table. The truth table for the copulative proposition $a \cdot b$ and its contradiction $-(a \cdot b)$ is as follows:

a	b	a·b	—(a·b)
T	T	T	F
T	F	F	T
F	T	F	T
F	F	F	T

Conditional, alternative, and disjunctive propositions express commitments to a sequential nexus between their elements. Hence they cannot be verified simply in terms of the truth values of their component propositions. Thus they are not truth-functional, and they cannot be expressed on a truth table manifesting all the conditions under which they will be determinately true or false. No combination of true or false premises guarantees the truth of $a \rightarrow b$, $a \leftrightarrow b$, $a \lor b$, $a \land b$, or $n - (a \cdot b)$. The only purpose of a truth table for these would be to indicate the situations in which they could not possibly be true. Although no combination of true and false components will guarantee the truth of $a \rightarrow b$, nevertheless, if a is true and b is false, $a \rightarrow b$ could not be true. Similarly $a \land b$ must be false if both a and b are true. The same is true for the other hypotheticals. Though no combination of truth values for their components guarantees their truth, some combinations do necessitate falsity. This is so because these combinations rule out the possibility that the sequential nexus which the proposition asserts between its components exists. Hence, truth tables cannot be used to verify or confirm hypotheticals, but they can be used to show when hypotheticals cannot be verified. With this understanding of copulative and hypothetical propositions, the following truth table expresses all the conditions for the verification and disconfirmation of copulative propositions as well as those which can be used to disconfirm impossible hypothetical propositions.[2]

a	b	a·b	a→b	a↔b	a∨b	a∧b	n — (a·b)
T	T	T	?	?	?	F	F
T	F	F	F	F	?	?	?
F	T	F	?	F	?	?	?
F	F	F	?	?	F	F	?

[2] Some logicians (both in the so-called Aristotelian tradition and in the school of mathematical logic) understand the verbal expressions of the alternative and disjunctive propositions as signs of a truth-functional proposition. They understand the inclusive alternative simply as an assertion that at least one of several component propositions is true (*note:* that at least one *is* true, not that, in the light of the others, one *must* be true). In similar fashion the exclusive alternative is understood simply as an assertion that at least one, but only one, of several component propositions *is* true. The disjunctive is taken as an assertion that at least one of several component propositions *is* false. Some logicians (ordinarily only those in the school of mathematical logic) take even the conditional expression as the sign of a truth-functional proposition. They understand the simple conditional simply as an assertion that it *is* not the case that the antecedent is true and the consequent is false. The reciprocal conditional is understood as an assertion that it *is* not the case that either component is true while the other is false. To speak of alternative, disjunctive, and conditional propositions in this sense is to use these terms equivocally. In recognition of this point, at least in reference to the conditionals, the type of implication understood for a conditional proposition taken as a truth-functional proposition is usually given a different name than the type of implication understood for a conditional taken as a strict hypothetical proposition. Truth-functional implication is called material implication, and strict implication is called formal implication. Actually truth-functional alternative, disjunctive, and conditional propositions are reductively of the copulative type (each one equivalently the contradiction of a more or less complex copulative proposition). Suppose we symbolize the truth-functional alternatives somewhat differently than the alternatives taken strictly. Adding an "f" for "factual," the truth-functional inclusive alternative can be symbolized a$\setminus^f/$b. The truth-functional exclusive alternative will be symbolized as a$/_f\setminus$b. a$\setminus^f/$b ≡ —(—a · —b), and a$/_f\setminus$b ≡ —(a · b) · —(—a · —b). Adding an "m" for "material," the truth-functional simple conditional can be symbolized as a$\xrightarrow{m}$b; and the truth-functional reciprocal conditional will be expressed as a$\xleftrightarrow{m}$b. a$\xrightarrow{m}$b ≡ —(a · —b), and a$\xleftrightarrow{m}$b ≡ —(a · —b) · —(—a · b). Finally, the truth-functional disjunctive is clearly nothing but the contradiction of an original copulative proposition, e.g., —(a · b). Truth-functional conditionals, alternatives and disjunctives, precisely as the copulatives, lend themselves completely to a truth table for purposes of verification and disconfirmation. The truth table for each of these is as follows:

a	b	a·b	a$\xrightarrow{m}$b	a$\xleftrightarrow{m}$b	a$\setminus^f/$b	a$/_f\setminus$b	—(a·b)
T	T	T	T	T	T	F	F
T	F	F	F	F	T	T	T
F	T	F	T	F	T	T	T
F	F	F	T	T	F	F	T

XII. A Schematic Résumé

In the past two chapters we have divided the proposition into its types through the use of logical division, codivision, and subdivision. Perhaps the following schematic outline will help to summarize the work of these two chapters:

```
                    ┌ Categorical  ┌ Simply attributive  ┌ Affirmative  ┌ Universal
                    │ (Simple)     │                     │              ┤ Particular
                    │              └ Modal               └ Negative     └ Singular
Proposition ┤
                    │              ┌ Conjunctive ┌ Copulative       ┌ Openly
                    │              │             │                  │ Occultly     ┌ Exclusive
                    │              │             └ Copulative plus  ┤ Adversative  ┤ Exceptive
                    └ Compound ┤                                    │              └ Reduplicative
                                   │                                └ Causal
                                   │              ┌ Conditional  ┌ Simple
                                   │              │              └ Reciprocal
                                   └ Hypothetical ┤ Alternative  ┌ Inclusive
                                                  │              └ Exclusive
                                                  └ Disjunctive
```

Each of the following propositions would be true if it were taken truth-functionally, but none of them is true if taken strictly as hypothetical: *Inclusive alternative* — Either Washington is the capital of the U.S.A. or Milwaukee is the home of the Yankees. Either Marquette University is in Wisconsin or the Civil War was fought in the seventeenth century. Either seven is an odd number or Communism is a threat to world peace. *Exclusive alternative* — Either Washington is the capital of the U.S.A. or Milwaukee is the home of the Yankees — but not both. Either seven is an odd number or the Civil War was fought in the nineteenth century — but not both. Abraham Lincoln was either the sixteenth U. S. president or the first man to swim the English Channel — but not both. *Disjunctive* — Washington is not both the capital of the U.S.A. and the home of the Braves. It is not true that seven is an odd number and that the Civil War was fought in the seventeenth century. St. Louis is not both south of New Orleans and a former site of the World's Fair. *Simple conditional* — If Washington is the capital of the U.S.A., then Milwaukee is the home of the Braves. If Washington is not the capital of the U.S.A., then Milwaukee is the home of the Braves. If Washington is not the capital of the U.S.A., then Milwaukee is not the home of the Braves. *Reciprocal conditional* — If and only if Washington is the capital of the U.S.A., then Milwaukee is the home of the Braves. If and only if Washington is not the capital of the U.S.A., then Milwaukee is not the home of the Braves. If and only if no man is rational, then every whole number is odd.

Exercise XII

1. Define: compound proposition, conjunctive proposition, copulative proposition, openly copulative proposition, occultly copulative (exponible) proposition, exclusive proposition, exceptive proposition, reduplicative proposition, adversative proposition, causal proposition, hypothetical proposition, simple conditional proposition, reciprocal conditional proposition, exclusive alternative proposition, inclusive alternative proposition, disjunctive proposition, truth-functional proposition, truth table.
2. What is meant by "sequential nexus," and how precisely does this serve as a principle of differentiation between types of compound proposition?
3. In what way are causal and adversative propositions like hypothetical propositions, and why, despite this, are they conjunctive rather than hypothetical?
4. What is the meaning of "antecedent" and "consequent" as used in reference to the conditional proposition?
5. Explain the primacy of the conditional among the hypotheticals, and illustrate the reduction of alternative and disjunctive propositions to conditional propositions.
6. Supply examples of each type of proposition so far considered, and evaluate each example as true or false with an explanation for this evaluation.
7. Classify each of the following propositions and, where the intelligible content is sufficiently familiar, indicate whether or not it is true. Give a determinate explanation for the falsity of those which are not true. Make a note of those which can be taken either as hypothetical or categorical.
    7.1. Only the good die young.
    7.2. A man who is afraid need not be a coward.
    7.3. Either logic is necessary for science or a course in logic is a waste of time.
    7.4. No animal except man has ever given evidence of moral responsibility.
    7.5. "In Flanders Fields the poppies blow
        Between the crosses, row on row,
            That mark our place, and in the sky
            The larks, still bravely singing, fly,
        Scarce heard amid the guns below."
            (Taken from "In Flanders Fields" by John McCrae.)
    7.6. Creatures are contingent beings, while the creator is absolutely necessary.
    7.7. Only men are rational.
    7.8. A Catholic politician cannot be both a good Catholic and a good politician.
    7.9. The man who has not known love is an unfulfilled man.

THE COMPOUND PROPOSITION   145

7.10. So long as a number is odd it cannot be divided by two without a remainder.
7.11. History cannot be counted on as an infallible index of the future.
7.12. Man is a corruptible being in so far as he is corporeal.
7.13. Logic is an art, since it is ordered to a work to be produced.
7.14. As long as, and only as long as, men have respect for God, the human race will be capable of a peaceful existence.
7.15. Not all negative propositions have universal predicates.
7.16. Cheating is immoral, for it is at least a violation against the virtue of truth-telling.
7.17. No nation can establish a manned station on the moon.
7.18. An argumentation is either valid or its conclusion does not follow from its premises.
7.19. A thing cannot be and not-be.
7.20. If and only if man is not free, the human race is not responsible for the chaotic condition of the world.
7.21. Logic is an art and not a science.
7.22. The Braves won the National League pennant in 1958 despite the fact that the Yankees won the American League pennant that same year.
7.23. If a man is moving from place to place, he must be walking.
7.24. Harvey Haddix is the only man ever to pitch twelve perfect innings in one major league ball game.
7.25. Either the Communists were very sharp or the Western leaders very naïve in the policy-making conferences held at the conclusion of World War II.
7.26. Nothing can be both a violation of human nature and something morally good for any man.
7.27. Either Communism will be ultimately unsuccessful or the free world will be destroyed.
7.28. All adult American citizens with the exception of the citizens of the District of Columbia are eligible to vote for the office of president of the United States.
7.29. Either golf is a healthy sport or lots of Americans are wasting their time.
7.30. A false modal proposition has a false mode or a false dictum.
7.31. Man is able to reproduce himself precisely in so far as he is an animal.
7.32. Either sound travels in waves or in particles, but not both.
7.33. Since every triangle is three-sided, a scalene triangle has no two sides equal to each other.
7.34. Causal propositions involve, as conjunctive, more than a commitment to a nexus between elements.

7.35. If and only if a subject of a proposition is a substance can its predicate be a genus.
7.36. If the present cold war becomes hot, the human race might exterminate itself.
7.37. If logic is a tool for the other sciences, then the other sciences naturally follow logic in the order of learning.
7.38. No contingent thing is the full explanation of its own existence.
7.39. A triangle is either scalene or equilateral, but not both.
7.40. All the predicables, with the exception of accident, are related essentially to their subjects.

8. What are the limits so far as the use of a truth table for hypotheticals strictly taken is concerned?
9. Defend the charge that to speak of truth-functional hypothetical propositions is to use the term "hypothetical proposition" equivocally.
10. Indicate T, F, or ? in the parentheses preceding each proposition after determining the truth value of the proposition to the extent to which this can be done in the light of the data given.

10.1. ( ) a · b, if a is true and b is true.
10.2. ( ) a $\overset{m}{\longrightarrow}$ —b, if a is true and b is false.
10.3. ( ) n — (a · b), if a is false and b is true.
10.4. ( ) —a $\wedge_?$ b, if a is true and b is true.
10.5. ( ) a↔b, if a is false and b is true.
10.6. ( ) —(—a · —b), if a is false and b is false.
10.7. ( ) a $\vee_?$ b, if a is true and b is true.
10.8. ( ) a $\wedge$ —b, if a is false and b is false.
10.9. ( ) —a→b, if a is true and b is true.
10.10. ( ) —a $\vee$ —b, if a is true and b is true.
10.11. ( ) —a $\overset{m}{\longrightarrow}$ —b, if a is true and b is true.
10.12. ( ) —a · —b, if a is false and b is false.
10.13. ( ) n — (a · b), if a is true and b is false.
10.14. ( ) —a $\wedge_?$ —b, if a is true and b is true.
10.15. ( ) a $\overset{m}{\longleftrightarrow}$ b, if a is false and b is true.

CHAPTER XIII

Relations Between Propositions

I. Propositional Relations

We have seen that terms can be logically related to one another in different ways. For example, *white* and *nonwhite* are related by way of contradiction, *illegible* is related to *legible* by way of privation, *animal* is the proximate genus of *man*, etc. Propositions can also be related to one another in different ways. We shall devote ourselves in this chapter to an investigation into several of the most significant logical relations which can be realized between propositions.

It is a property of propositions to be either true or false. The truth-status of a proposition logically related to another is ordinarily an index of the truth-status of that other proposition. For example, we find that *Every man is potentially an artist* and *No man is potentially an artist* are related as contraries. We shall see that contraries cannot be true together. Thus, from the truth of either of these two propositions we can move immediately to the falsity of the other. Many logicians speak of the immediate passage from the truth (or falsity) of one proposition to the truth (or falsity) of a logically related proposition as an instance of immediate inference. "Inference," however, properly refers to the rational movement of the intellect in its third operation; and, as we shall see, to move from the truth-status of one proposition to that of an immediately related proposition is not to reason. Consequently, it is inaccurate to speak here of an immediate inference. It is better to say simply that we are concerned with propositions which are determinately related to one another and which are, because they are so related, indices for the truth-status of each other.

We have discussed both categorical and compound propositions and their subdivisions. In this chapter we will limit ourselves to a consideration of the relations between categorical propositions, and, to be even more restrictive, only of relations between categorical propositions which are simply attributive rather than modal. Not that there are no logically significant relationships between modal propositions and between compound propositions of one type or another. As a matter of fact, many of the propositional relations we shall investigate within the context of the simply attributive categorical proposition are also found between modal propositions and between compound propositions. The reason for restricting our consideration is simply to limit our investigations prudently within the bounds indicated by the relatively short time available for a semester's study. Since simply attributive categorical propositions are the fundamental type of proposition, we shall be able, despite the restrictions in the breadth of our consideration, to acquire a solidly basic grounding in the theory of propositional relations.

We have spoken of six types of categorical proposition, the A, E, I, O, A^1, and E^1 propositions. Each type admits of different propositional relations. For the most part the logical relations belonging to the universal and particular propositions offer relatively little difficulty. However, the relations belonging to singular propositions frequently involve difficulties which make their investigation relatively troublesome. But the student may be encouraged to learn that singular propositions are not nearly so significant logically as the others. Logic is a tool pre-eminently aimed at science, and singular propositions are for the most part scientifically irrelevant. Thus the beginning student can put aside the difficulties involved in considering the logical relationships of singular propositions in order to concentrate more on those belonging to universal and particular propositions. With this in mind this chapter has been planned so that the various propositional relations are discussed first of all in reference to A, E, I, and O propositions. A separate section of the chapter follows in which all the propositional relations considered in the chapter are discussed in the context of the A^1 and E^1 propositions.

II. The Nature and Types of Opposition

Opposition is the logical relationship between propositions which affirm and deny the same predicate of the same subject. Three different modes of opposition can exist between propositions, namely, *contradiction, contrariety,* and *subcontrariety*.

1. Contradiction exists between two propositions with the same subject and predicate which purely and simply deny one another. Propositions *purely and simply* deny one another when the truth of either excludes the truth of the other while the falsity of either excludes the falsity of the other. The contradictory of a given proposition, say, *Every man is naturally logical*, is equivalent to an assertion that this proposition is not true,

that is, *It is not true that every man is naturally logical.* But this is asserted in *Some man is not naturally logical.* Thus, *Some man is not naturally logical* and *Every man is naturally logical* are related by way of contradiction. They purely and simply deny one another: if one is true the other is false, and if one is false the other must be true. Contradictory opposition exists between both A and O propositions with the same subject and predicate and E and I propositions. An example of contradictory A and O propositions is: *All men are white* (A); *Some man is not white* (O). An example of contradictory E and I propositions is: *No animals are rational* (E); *Some animal is rational* (I). Another example of A-O contradictories is: *All men are rational* (A); *All men are not rational* (O). Although this O proposition may seem to be an instance of an E proposition, it is, as we have seen, particular. As long as one of these pairs of propositions is true the other is false and as long as one is false the other is true.

2. *Contrariety* is the mode of opposition between two propositions with the same subject and predicate which deny not only one another but even less extended formulations of each other.

An A proposition with a given subject and predicate can be considered a more extended formulation of an I proposition with the same subject and predicate. The same thing, of course, is true for an E proposition and a corresponding O. An A proposition, say, *Every criminal is neurotic*, is a denial of the corresponding E with the same subject and predicate, namely, *No criminal is neurotic*. It is at the same time a denial of (in fact, the contradiction of) the O proposition which represents a less extended formulation of the E, namely, *Some criminal is not neurotic.* Because the A in question is a denial of the corresponding E, and at the same time a denial of the less extended formulation of that E, it is said to be related to the E by way of contrariety. Just as the A proposition is the contrary of the E, the E, in turn, is the contrary of the A, for the E denies the A and also the I which is the less extended formulation of that A.

At first glance contrariety might seem to be a stricter mode of opposition than contradiction. However, contrariety is less strict than contradiction. Like contradictories, contraries cannot both be true, because they deny one another. Unlike contradictories, however, they can both be false, because they not only deny one another but also less extended formulations of each other. This additional denial actually introduces a weakness into the opposition of contrariety so that contraries need not be opposed in falsity. *No man is neurotic* and *Every man is neurotic* are contrary propositions. Since they deny one another they cannot both be true. Yet both happen to be false, for some men are neurotic and some men are not neurotic. It is, of course, not always the case that contraries are both false. *Every animal is sentient* and *No animal is sentient* are contraries, and the first is true while the second is false. The rule concerning contraries is simply this: since contraries cannot be true together but may be false together, if one is true the other must be false, but if one is false the other is questionable in the light of the first. A person may know that the second proposition is true or false while the first is false, but he will not know this on the evidence of the falsity of its contrary. He will know it only on other grounds. However, if he knows, to begin with, that the first proposition is true, then the fact that the second is the contrary of the first is sufficient evidence for him to assert its falsity. Although contraries might both be false, they can never both be true.

Before proceeding from contrariety to the third type of opposition it is worthwhile to suggest the following rule as a practical canon of effective debating: never propose a contrary when a contradiction will do. It is true that one can show that his opponent's position is false if he can establish the contrary of his opponent's position. Yet we must remember that contrary positions may both be false. If one opposes a false position with the equally false contrary of this position one does not destroy the opposition and, in fact, one may seem to strengthen it. The surest and simplest destruction of a position expressed either in an A or an E proposition is by establishing the contradictory O or I proposition. How much simpler to refute the charge that no Americans are loyal by presenting in evidence some one loyal American than by attempting to establish the loyalty of every American.

3. Subcontrariety arises between two propositions with the same subject and predicate which do not deny one another but which do deny the more extended formulation of the other.

Because *I* and *O* propositions each have a particular subject, an *I* proposition does not deny the corresponding *O*; nor, of course, does the *O* deny the *I*. However, the *I* does deny (by way of contradiction) the *E*, which is the more extended formulation of the *O*, and the *O* denies the *A*, which is the more extended formulation of the *I*. Thus the *O-I* propositions with the same subject and predicate are subcontraries. Subcontrariety is a mode of *opposition* in only a tenuous sense, for subcontraries may both be true since they do not deny one another. Thus, *Some horses are spotted* and *Some horses are not spotted* are legitimately subcontraries, and each is a true proposition. Of course, subcontraries need not both be true. *Some men are rational* is true and, because all men are rational, *Some men are not rational* is false. Whatever opposition there is in subcontrariety is seen in reference to falsity, not truth. Though subcontraries might both be true they can never both be false. Suppose *Some A is B* is false, then *Some A is not B* must be true. The reason for this can be shown in the following way. If *Some A is B* is false, then its contradictory *No A is B* is true, since contradictories cannot be true together. If, at the same time, *Some A is not B* were false, then, its contradictory *Every A is B* would be true. But if that were the case then *No A is B* and *Every A is B* would both be true; but this is impossible because they are contraries. Thus, if the *I* (*Some A is B* in our example) is false, the corresponding *O* (*Some A is not B*) cannot be false. A similar argument can be used to show that the *I* must be true if the *O* is taken as false.

The following examples of opposed propositions should help familiarize the student with the modes of opposition:

1. Every democracy is a politically sound nation. (*A*)
   Some democracy is not a politically sound nation. (*O*)
   (Contradiction)

2. Some materialists are Marxists. (*I*)
   Some materialists are not Marxists. (*O*)
   (Subcontrariety)

3. All that glitters is gold. (*A*)
   All that glitters is not gold. (*O*)
   (Contradiction)

4. No beings-of-the-reason can be categorized. (*E*)
   Some beings-of-the-reason can be categorized. (*I*)
   (Contradiction)

5. Some of the men who were not at the scene were unable to recall the event. (*I*)
   Some of the men who were not at the scene were not unable to recall the event. (*O*)
   (Subcontrariety)

6. None of the players was tired. (*E*)
   All of the players were tired. (*A*)
   (Contrariety)

7. All the troops crossed the Delaware. (*A*)
   Some of the troops did not cross the Delaware. (*O*)
   (Contradiction)

8. No non-A is B. (E)
   Some non-A is B. (I)
   (Contradiction)

9. Some A is not non-B. (O)
   Some A is non-B. (I)
   (Subcontrariety)

10. Every A is non-B. (A)
    Every A is not non-B. (O)
    (Contradiction)

III. Subalternation

Propositions related by way of opposition always involve the same subjects and predicates and different quality, i.e., affirmation or negation. Sometimes they involve different quantity, sometimes the same quantity. What about propositions which have the same subject and predicate and the same quality and which differ only in quantity? These propositions are surely related to one another, but they are just as surely not opposed to one another. We have already spoken of A and I (and E and O) propositions with the same subject and predicate. We said that the universal proposition was related to the particular as a more extended formulation to a less extended formulation. The universal and particular propositions are related to one another analogously as superiors and inferiors. The universal proposition includes the corresponding particular proposition as a logical whole contains one of its subject parts. This relationship is called *subalternation*, and the superior proposition is called the *subalternand*, the inferior the *subalternate*. A predicate can be said universally of a logical subject only if it can be said of every instance of that subject. Hence the subalternand cannot be true unless the subalternate is also true. But there is no guarantee that a predicate can be said universally of a subject just because it can be said of some instances of this subject. Thus the truth of the subalternate leaves undetermined the truth of the subalternand. This implies, of course, that the subalternand must be false if the subalternate is false, but it allows for the truth or falsity of the subalternate in the light of a false subalternand. Hence, if *Every triangle is three-sided* is true, *Some triangle is three-sided* must also be true; and if *Some men are not rational* is false, *No men are rational* must also be false. But the falsity of *No man is a Communist* leaves the truth-status of *Some man is not a Communist* in question; and the truth of *Some horse is spotted* leaves the truth-status of *Every horse is spotted* in question.

The following are examples of propositions related by subalternation:

1. Every Marine is taught to respect his rifle. (A)
   Some Marine is taught to respect his rifle. (I)
   (Superior [= subalternand] to inferior [= subalternate])

2. Some horses are not spotted. (O)
   No horses are spotted. (E)
   (Inferior to superior)

3. No unintelligent man is fit to lead the nation. (E)
   Not every unintelligent man is fit to lead the nation. (O)
   (Superior to inferior)

4. Some non-A is non-B. (I)
   Every non-A is non-B. (A)
   (Inferior to superior)

5. Every A is not B. (O)
   No A is B. (E)
   (Inferior to superior)

Before getting further into the discussion of propositional relations it is necessary to insist upon a point which applies not only to the different modes of opposition but to any relationship between propositions. One of the reasons why propositions are logically related to one another is that they involve *identical* terms. Thus contraries have the same subject and predicate while differing in quality but not in quantity. For terms to be identical they must, as we have seen, involve generally the same signification and supposition. It is understood throughout this chapter whenever we speak of moving from the truth-status of one proposition to that of a related proposition that there is no logically disruptive break in the signification or the supposition of any given term from proposition to proposition. Let us illustrate by way of reference to the relation of subalternation. We have said it is legitimate to go from the truth of a subalternand to the truth of its subalternate, that is, from an A to its corresponding I and from an E to its corresponding O proposition. This would not be true if the A proposition expressed only the possible existence of its subject and the I its actual existence. It can be true only if the A as well as the I expresses actual existence or if the I as well as the A expresses only possible existence. Furthermore, although an A may frequently express only possible existence it can intend actual existence;

and although an *I* frequently expresses actual existence, it may intend only possible existence. The point here is that the terms must be the same from proposition to proposition, and this means that they involve the same signification and at least generally the same supposition.

IV. The Square of Opposition

Logicians have traditionally employed a visual aid, the "square of opposition," to assist in the handling of propositions involving opposition or subalternation. The "square" is a rectangular figure whose corners represent respectively *A*, *E*, *I*, and *O* propositions and whose sides and diagonals as well represent the three modes of opposition and subalternation. It serves as a graphic reminder that *A* and *O* (and *E* and *I*), propositions with the same subject and predicate are related by way of contradiction, that *A* and *E* propositions are contraries of one another, and that *I* and *O* propositions are subcontraries of one another. Finally it is a reminder that the *A* is related to the *I* and the *E* to the *O* as superiors to inferiors in the relationship of subalternation.

THE SQUARE OF OPPOSITION

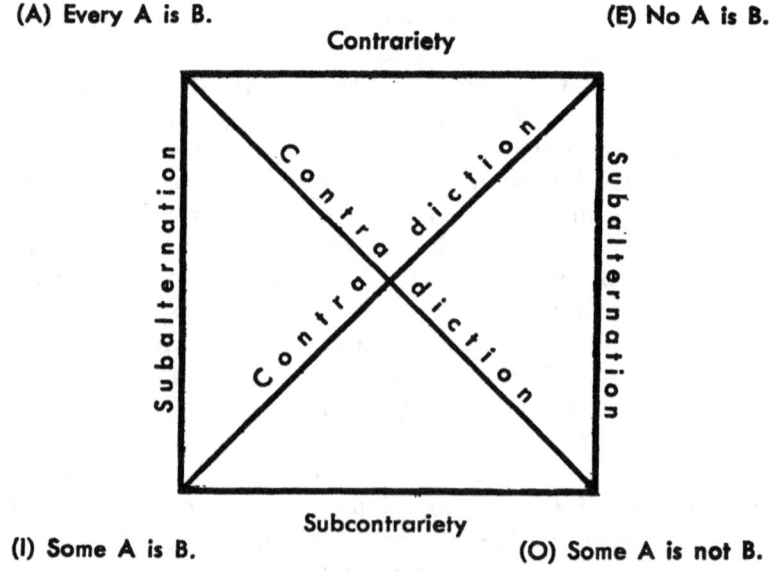

It may be helpful to add a truth table as a companion piece to the "square." Given the truth value of one proposition on the "square," the

RELATIONS BETWEEN PROPOSITIONS   155

truth table expresses graphically the truth values of each of the others. The four horizontal columns in the table represent the A, E, I, and O propositions. There are eight vertical columns in the table, each of which records a truth value which is assumed for one of the four propositions, along with the correlative truth values of the others which follow from this. In each case the truth value which is assumed as a standard is indicated by being encircled. Begin each time with the encircled T or F proposition to ascertain the truth-status of the other propositions in the same horizontal columns.

TRUTH TABLE FOR THE SQUARE OF OPPOSITION

(A)	Every A is B.	(T)	F	?	F	(F)	?	F	T
(E)	No A is B.	F	(T)	F	?	?	(F)	T	F
(I)	Some A is B.	T	F	(T)	?	?	T	(F)	T
(O)	Some A is not B.	F	T	?	(T)	T	?	T	(F)

Neither the "square" nor its truth table are suggested as mechanical substitutes for an understanding of the relationships of opposition and subalternation and the logical rules which are determined by these relationships. The "square" is recommended as an aid in seeing the concrete realization of logical relationships in determinate propositions. The truth table can be used as a check against the erroneous manipulation of propositions related by way of opposition or subalternation. Perhaps the most profitable use of the truth table for the student consists first of all in appreciating the principles according to which it is constructed and then in repeated reconstructions of it, not from memory, but rather from an understanding of the principles underlying it.

V. The Nature and Types of Conversion

Conversion is the logical relationship between propositions with inverted or interchanged extremes which express either totally or partially the same truth. The act of converting is achieved by moving from an original proposition (called the convertend) by inverting its subject and predicate in a second proposition (called the converse) which expresses the same truth as the original. The quality of the convertend and its converse must always be the same. Some propositions can be legitimately converted simply by interchanging their subjects and predicates while leaving the quantity unchanged. This is called *simple conversion*. Other propositions can be legitimately converted only if there is a change in the quantity of the original proposition. This is called *accidental conversion*. *Some Americans are neurotic* can be simply converted to *Some neurotics are American*. *Every Hoosier is an American* cannot be simply converted to *Every American is a Hoosier*, but it can be accidentally converted to *Some Americans are Hoosiers*. To be legitimate a simple converse proposition must express totally the same truth expressed by the convertend. To be legitimate an accidental converse proposition must express partially the same truth expressed by the convertend. A converse expresses totally the truth of the convertend whenever both of its terms have the same extension that they had in the convertend. The converse only partially expresses the truth of the convertend whenever either of its terms has less extension than it had in the convertend. (An attempted converse expresses something not expressed by the convertend whenever any term in the converse is more extended than the same term is in the convertend.) *Some neurotics are American* is a legitimate simple converse of *Some Americans are neurotic* because both terms have particular extension in both the convertend and the converse. *Every American is a Hoosier* is an illegitimate converse for *Every Hoosier is an American* because American is taken with universal extension in the converse and with only particular extension in the convertend. However, *Some American is a Hoosier* is a legitimate accidental converse for *Every Hoosier is an American*. American is taken with the same extension in both convertend and converse, while *Hoosier* is taken with less extension in the converse than in the convertend. If any of the terms in an attempted converse proposition should have more extension than it had in the convertend, then the conversion is illegitimate. That is why a proposition such as *Some professors are not scientists* cannot be converted. In any attempted converse *professor* would be the predicate of a negative proposition. As such, it would be universal and therefore more extended than it was in the convertend, where it was taken particularly.

The legitimate simple converse expresses exactly the same truth as its convertend. Hence, if the convertend is true the converse is true, and vice versa. If the convertend is false the converse is false, and vice versa. The legitimate accidental converse expresses partially what its convertend expresses. Since this relation is the same as that between subalternand and subalternate propositions the rules that apply for subalternation apply for accidental conversion. Hence, a true convertend implies a true converse, but not vice versa; and a false converse implies a false convertend, but not vice versa.

Conversion is of importance both on the level of the second operation of the intellect and on the level of the third operation. But we should note that the movement from a convertend to its legitimate converse is not reasoning. The converse is a new proposition, but it does not represent a new truth. The act of converting is an *immediate* passage from a truth expressed in one way to that *same* truth expressed differently. This, it should be noted, is generally the case for the operations discussed in this chapter. None of them is an instance of reasoning. The act of reasoning is a *mediate* movement from previous knowledge to new knowledge which is potentially in the previous knowledge. Still the converse is a *different expression* of the same truth. It is logically worthwhile to be able to express a given truth in as many ways as possible and especially to be able to recognize the same truth no matter how it is presented. It is possible that, for a given person, one expression of a truth might be more meaningful than another. When we study the categorical syllogism we shall see another purpose for conversion. We will find that syllogisms in one form can (and sometimes should) be reduced to syllogisms in another form. This usually is done by converting the propositions of the original syllogism.

VI. Legitimate Conversions

Since some propositions can be simply converted, some only accidentally converted, and some not at all, we must determine exactly the number and nature of the legitimate conversions. Since A, E, I, and O propositions all theoretically admit two converses, the simple and the accidental, there are eight different converses which we can test to see if they are legitimate. As we have seen, those which involve terms which have greater extension than they had in the convertend must be rejected as illegitimate. The results of the investigation into the legitimacy of the eight possible converses can be expressed as follows:

(A) Every A is B.	Simple	Every B̶ is A̲.
	Accidental	Some B is A̲.
(E) No A is B.	Simple	No B̄ is Ā.
	Accidental	Some B is not Ā.
(I) Some A is B.	Simple	Some B is Ā.
	Accidental	Every B̶ is Ā.
(O) Some A is not B.	Simple	Some B is not B̶.
	Accidental	No B̄ is B̶.

Whenever a term in the converse has the same extension that it had in the convertend, a line is drawn over it. When it has less extension, a line is drawn under it. When it has more extension it has been crossed out. Any converse with a term which has been crossed out is illegitimate. Those with lines over both terms are legitimate simple converses, as they express the whole of the original truth. Legitimate converses with any term underlined are accidental converses, as they express only a part of the original truth. There are two legitimate simple converses, two legitimate accidental converses. Only the E and the I propositions can be simply converted. The E can also be accidentally converted, along with the A. The O can be converted neither simply nor accidentally.

We have ruled out the possibility of the simple conversion of an A proposition because its predicate, which is particular, would have to be universal as subject of the converse. This is an indisputable point so long as we are considering the A proposition from the formal point of view. Formally the A proposition can guarantee no more than the possibility of an accidental converse. However, in A propositions which happen to have predicates which are coextensive with their subjects simple conversion is legitimate. This exception is made *strictly on the basis of the matter of the proposition*. The *form* of an A proposition never guarantees a legitimate simple conversion. Thus, A propositions whose predicates are related to their subjects as definition, specific difference, or specific property to species can be simply converted so long as the conversion is made in virtue of the fact that the predicate happens to be coextensive with its subject. In this way I can convert *Every man is a rational animal* to *Every rational animal is a man* and *Every man is capable of speech* to *Every thing capable of speech is a man*. The predicate of the first is the definition of the subject, and the predicate of the second is a specific property of the subject.

The following examples of legitimate conversion should help to familiarize the student with both simple and accidental conversion:

1. No triangles are four sided figures. (E)
No four-sided figures are triangles. (E)
(Simple conversion)
2. Every logical relationship is a being-of-the-reason. (A)
Some being-of-the-reason is a logical relationship. (I)
(Accidental conversion)
3. No coward is able to be a true hero. (E)
Not everyone able to be a true hero is a coward. (O)
(Accidental conversion)

4. Some whole numbers are even. (*I*)
   Some even things are whole numbers. (*I*)
   (Simple conversion)
5. Every genus is an example of a predicable. (*A*)
   Some example of a predicable is a genus. (*I*)
   (Accidental conversion)
6. No scientist can afford to scorn logic. (*E*)
   None who can afford to scorn logic are scientists. (*E*)
   (Simple conversion)
7. Bats have gnats. (*A*)
   Some who have gnats are bats. (*I*)
   (Accidental conversion)
8. No non-*A* is *B*. (*E*)
   No *B* is non-*A*. (*E*)
   (Simple conversion)
9. Some *A* is non-*B*. (*I*)
   Some non-*B* is *A*. (*I*)
   (Simple conversion)
10. No *A* is non-*B*. (*E*)
    Some non-*B* is not *A*. (*O*)
    (Accidental conversion)

VII. Obversion

Obversion is the logical relationship between propositions which express the same truth by affirming and denying contradictory predicates of the same subject. Obversion, unlike conversion, is valid for all types of propositions. An original proposition (called the obvertend) can be obverted simply by changing its quality and contradicting its predicate. The result is a second proposition (called the obverse) which expresses the same truth as the original. This is so because the two propositions have opposite quality and predicates which are contradictorily opposed to one another. Contradictory terms are related to one another as being and non-being, or as affirmation and negation. Accordingly when one of two contradictory predicates is affirmed of a subject while the other is denied of the same subject, the resulting propositions say exactly the same thing, but, of course, in different ways. Since the obverse and its obvertend express exactly the same truth, if either is true the other is true, and if either is false the other is false.

Strictly speaking, the predicate of the obverse should be the contradiction of the predicate of the obvertend. However, obversion is legitimate as long as the predicates function as contradictories, even if, strictly speaking, they are not. This is the case with predicates which are opposed as immediate contraries or with predicates of which one is the privation of the other. Here, however, we must use these terms only within the restricted area of their relevance. Strict obversion is based on the fact that one term can be predicated of a subject if its contradictory cannot. This is not absolutely true either for immediate contraries or for privatives. However, *within the proper genus of things*, the immediate contrary of a term can be said of that of which the term itself cannot be said, and vice versa. Thus, if a whole number is not odd, that whole number is even. And if a whole number is even, that whole number is not odd. In similar fashion, the privative of a term can be predicated of that of which the term itself cannot, and vice versa, so long as the terms are used *with reference to the proper genus of things*. If a man is not able to see, that man is blind; and if a man is blind, that man is not able to see. But we could not say that Fido is even because Fido is not odd, nor that Plymouth Rock is blind because Plymouth Rock is not able to see. The reason is simply this: dogs and rocks are not properly within the genera of things for which *odd* or *even* and *able to see* or *blind* are relevant predicates.

Obversion is, in general, logically significant for the same reason that conversion is important. It is worthwhile to be able to express the same thing in more than one way. Obversion is especially suited to the elimination of excessive negatives in the expression of a truth. A proposition overly laden with negative elements is often hard to grasp. Its obverse will eliminate all or some of the negative elements and may help greatly to clarify the meaning of the original proposition. For example, *Some of Poe's short stories are not nonexciting* is more clearly expressed as *Some of Poe's short stories are exciting*. Similarly, *None of the benefits of the policy are nontransferable* is more clearly expressed as *All of the benefits of the policy are transferable*.

Examples of obversion include:
1. **Every man is rational. (A)**
   No man is nonrational. (E)
2. **Some solids have a density less than that of water. (I)**
   Some solids are not things which do not have a density less than that of water. (O)
3. **Some men are illiterate. (I)**
   Some men are not literate. (O)

4. Some politicians are not trustworthy. (O)
   Some politicians are untrustworthy. (I)
5. Every horse is not spotted. (O)
   Some horse is nonspotted. (I)
6. No animals are nonsentient. (E)
   Every animal is sentient. (A)
7. Some A is non-B. (I)
   Some A is not B. (O)
8. Every A is B. (A)
   No A is non-B. (E)
9. Every A is not non-B. (O)
   Some A is B. (I)
10. No A is non-B. (E)
    Every A is B. (A)

VIII. Contraposition

Neither an A nor an O proposition admits of simple conversion. However, logicians have discovered a substitute for simple conversion for both the A and O propositions. It is a complex relationship spoken of as *contraposition*. Contraposition is complex because it involves a combination of obversion and conversion. An original A or O proposition cannot be simply converted, but each can be obverted. The obverse of an A is an E, and the obverse of an O is an I. The E and I propositions can be simply converted, and their converses can be obverted once again. The obverse of the simple converse of the obverse of an original proposition is called the contrapositive of that original proposition. The original proposition is called the contraponend, and the relationship between the two propositions is the complex logical relationship of contraposition. Neither the E nor the I proposition can be contraposed, for in each case the obverse of the original is a proposition which defies simple conversion. Since the contrapositive is the obverse of the simple converse of the obverse of the contraponend, it expresses exactly the same truth as the contraponend. Accordingly, if either of the propositions related by way of contraposition is true, the other must be true. If either is false, the other must be false. Contraposition can be effected simply by inverting the extremes of the contraponend and by contradicting them, leaving their quantity and quality unchanged. *All nurses are noncombatants* can be contraposed by contradicting its subject and its predicate and by using them in inverted order in an A proposition. The resulting proposition is *All combatants are nonnurses*. In order to see that this is the legitimate contrapositive of *All nurses are noncombatants*, let us contrapose step by step:

All nurses are noncombatants. (A)   (The original proposition)
No nurses are combatants. (E)   (The obverse of the original)
No combatants are nurses. (E)   (The simple converse of the obverse of the original)
All combatants are nonnurses. (A)   (The obverse of the simple converse of the obverse of the original)

Some A is not B is an O proposition. The O proposition whose predicate is the contradictory of the subject of the original proposition and whose subject is the contradictory of its predicate is *Some non-B is not non-A*. Let us see that this is the legitimate contrapositive of *Some A is not B* by contraposing step by step:

Some A is not B. (O)   (The original proposition)
Some A is non-B. (I)   (The obverse of the original proposition)
Some non-B is A. (I)   (The simple converse of the obverse of the original proposition)
Some non-B is not non-A. (O)   (The obverse of the simple converse of the obverse of the original proposition)

Contraposition, as we have seen, is not a simple logical relationship. It is complex, involving two obversions and one conversion. There is, practically speaking, no limit to the number of complex relationships which can be set up between propositions. Contraposition is considered separately and has been given a special name because it can be done very simply and because the contrapositive serves as a convenient substitute for a simple converse for propositions which can have no simple converse. It is a helpful exercise for students to relate various propositions by way of different complex relationships. For example, *No B is non-A* is the obverse of the superior of the simple converse of *Some A is B*. *No B is non-A* might also be considered the superior of the obverse of the simple converse of *Some A is B*. In either case it can be seen that if *No B is non-A* is true, *Some A is B* is true; and that if *Some A is B* is false, *No B is non-A* must be false.

The following examples of contraposition will help to familiarize the student with this complex logical relationship:

1. Some nonathletes are not healthy. (O)
   Some who are nonhealthy are not athletes. (O)
2. Not every non-Catholic is one who fears the Vatican. (O)
   Some who do not fear the Vatican are not Catholic. (O)
3. Everything nonsentient is nonanimal. (A)
   Every animal is sentient. (A)
4. Some A is not non-B. (O)
   Some B is not non-A. (O)
5. Every non-A is B. (A)
   Every non-B is A. (A)

IX. Propositional Relations for the Singular Proposition

Singular propositions do not admit of opposition, subalternation, conversion, and obversion as readily and as clearly as do universal and particular propositions. Here we shall consider each of the propositional relations investigated in this chapter in the context of the A^1 and E^1 propositions. This consideration will reveal many difficulties not encountered with the A, E, I, and O propositions.

1. Contradiction

Here there is little difficulty. A^1 and E^1 propositions do admit of contradiction. As a matter of fact, the contradiction of the A^1 is the E^1 with the same subject and predicate, and, of course, the contradiction of the E^1 is the corresponding A^1. *This figure is not a triangle* is the contradiction of *This figure is a triangle. John is a sick man* is the contradiction of *John is not a sick man.*

2. Contrariety

Singular propositions can never be contraries; for, as singular, they cannot deny another proposition *as well as* a less extended formulation of that other proposition.

3. Subcontrariety

Strictly speaking subcontrariety is a logical relation exclusively possessed by *I* and *O* propositions. Nevertheless there is a relationship between the E^1 proposition and its corresponding *I*, and between the A^1 and its corresponding *O*, which is similar to the relation of subcontrariety. The A proposition cannot properly be said to be a more extended formulation of the A^1 in exactly the way in which it is a more extended formulation of the I. Given the A, the I is known immediately to be true, but, given the A, the A^1 is known to be true only after the existence of its singular subject is known. Nevertheless, once the existence of the singular subject of the A^1 is presumed, the A^1 can be seen to fall under the A in a manner analogous to the way the I falls under it. The same, of course, is true for the E and E^1 propositions. The I is the subcontrary of the O, and vice versa, precisely because each denies the universal proposition under which the other falls, even though they do not deny one another. Likewise the I does not deny the E^1, but it does deny the E under which the E^1 analogously falls; and the O does not deny the A^1, but it does deny the A under which the A^1 analogously falls. Thus the I and the E^1, as well as the O and the A^1, are related in a fashion similar to the way in which subcontraries are related to one another.

Hence as long as the singular subject of the A^1 proposition is presumed to exist and to be under consideration, the falsity of the A^1 implies the truth of the corresponding O, and the falsity of the O implies the truth of the corresponding A^1. Nothing follows from the truth of either. For the same reason, what has been said of the A^1 and the O can also be said of the E^1 and the I.

Thus, if it is false that some man is not a pure spirit, it must be true that this man is a pure spirit, so long as this man is understood to exist as a man. In similar fashion, if it is false that some Communist is a good Christian, it must be true that this Communist is not a good Christian, so long as this Communist is understood to exist as a Communist. Further, if it is false that this fish is a salmon, it must be true that some fish is not a salmon, and if it is false that this man is not a scientist, it must be true that some man is a scientist.

4. Subalternation

The case of subalternation and the singular proposition is like that of subcontrariety. In the strict sense singular propositions do not admit of subalternation. Yet they do admit of a relationship similar to that of subalternation. As a matter of fact, they can be thought of as related to the universal in a manner analogous to the way its inferior is related to it. Similarly they are related to the particular in a manner analogous to the way its superior is related to it. By subalternation the truth of the subalternand implies the truth of the subalternate, but not vice versa. The A is related to the A^1, and the E to the E^1, in such fashion that if the universal is true, the singular proposition must be true as long as its singular subject is presumed to exist. Because of the relation between the A and A^1 and between the E and E^1 propositions, if either of the singular propositions is false, the corresponding universal is false also. The A^1 is related to the I, and the E^1 to the O, so that if the singular proposition is true, the corresponding particular is true and if the particular proposition is false, any corresponding singular proposition must also be false.

Thus, if it is true that every saint is a hero, it must be true that *this* saint is a hero, so long as this saint is admitted to exist as a saint. And if it is true that this saint is a hero, surely it is true that some saint is a hero. Similarly, if it is true that no plant is sentient, it must be true that this plant is not sentient, so long as this plant is admitted to exist as a plant. And if this plant is not sentient, certainly some plant is not sentient. Further, if it is false that some parrots can talk, it must be false that this parrot can talk. And if it is false that this parrot can talk, it must be false that every parrot can talk. If it is false that some plants are not corruptible, it must be false that this plant is not corruptible. And if it is false that this plant is not corruptible, it must be false that no plant is corruptible.

5. Simple Conversion

Conversion is intended to restate the truth of a proposition differently. When the original proposition is singular this can be done only if the singularity of the subject of the convertend is retained in the predicate of the converse. The legitimate converse of a singular proposition must, then, have a singular predicate. This is an unusual logical situation. However, it can be tolerated. It is possible to convert both A^1 and E^1 propositions simply, so long as the singularity of the subject of the convertend is retained in the predicate of the converse. The A^1 is converted to an I proposition with a singular predicate. The E^1 is converted to an E proposition with a singular predicate. *Bertrand Russell is a well-known British philosopher* is converted to *Some well-known British philosopher is Bertrand Russell*. *John is not a hero* is converted to *No hero is John* (*this* individual John).

6. Accidental Conversion

As long as the singularity of the subject of the convertend is retained in the predicate of the converse, an E^1 proposition can be accidentally as well as simply converted. *John is not a hero* is accidentally converted to *Some hero is not John.*

7. Obversion

There is no difficulty in obverting A^1 and E^1 propositions. They obvert as easily as, and in exactly the same way as, universal and particular propositions. *The cold war is a reality* is obverted to *The cold war is not a nonreality. The cold war is not a fiction* is obverted to *The cold war is a nonfiction.*

8. Contraposition

Singular propositions admit of complex relations. For example, *John is not nonjust* is the obverse of the contradiction of *John is not just*. However, because of the unusual status of the predicate for the converse of a singular proposition, most complex relations which involve conversion are of questionable value for singular propositions. This is the situation as far as contraposition is concerned.

X. Résumé

It may be helpful here at the end of this chapter to summarize these rules in schematic fashion. Once again the student is warned against a completely mechanical use of the schema.

Opposition
    Contradiction   T↔F
    Contrariety   T→F   F→?
    Subcontrariety   T→?   F→T

Subalternation
    Subalternand to Subalternate   T→T   F→?
    Subalternate to Subalternand   T→?   F→F

Conversion
    Simple conversion   T↔T   F↔F
    Accidental conversion   T→T   F→?
          ?←T   F←F

Obversion   T↔T   F↔F
Contraposition   T↔T   F↔F

There are many helpful logical exercises which can be recommended for the matter treated in this chapter. One good exercise is to construct a number of different propositions and then attempt to make as many logically related propositions as possible for each of these. The exercise is completed by assuming first a true status, and then a false status, for the original proposition and by attempting to ascertain the consequent truth-status of each of its related propositions. As a practical hint the student is reminded that it is usually prudent to state propositions in strictly logical form before attempting to manipulate them. Suppose, for example, *Some Russians are Communist* is taken for a start. This is an *I* proposition, and it is already in strictly logical form. An *I* proposition has a contradiction, a subcontrary, a superior, a simple converse, and an obverse. If it is true, its contradiction is false, its subcontrary is questionable, its superior is questionable, its simple converse is true, and its obverse is true. If it is false, its contradiction is true, its subcontrary is true, its superior is false, its simple converse is false, and its obverse is false. Thus, beginning with *Some Russians are Communist*, the following exercise is called for:

Original proposition	— *Some Russians are Communist.*	T	F
Contradiction	— *No Russians are Communist.*	F	T
Subcontrary	— *Some Russians are not Communist.*	?	T
Superior	— *Every Russian is Communist.*	?	F
Simple Converse	— *Some Communists are Russians.*	T	F
Obverse	— *Some Russians are not non-Communist.*	T	F

Exercise XIII

1. Define: propositional relation, inference, opposition, contradiction, contrariety, subcontrariety, subalternation, subalternand (superior), subalternate (inferior), conversion, simple conversion, accidental conversion, convertend, converse, obversion, obvertend, obverse, contraposition, contraponend, contrapositive.
2. Generally compare relations between propositions to relations between terms.
3. Distinguish between a pure and simple denial and any other type of denial.
4. Why is contradiction a stricter mode of opposition than contrariety is?
5. Why is subcontrariety an instance of opposition taken only in a wide sense?
6. Why must contradictories have opposite truth values? Why can't contraries be true together? Why can contraries be false together? Why can subcontraries be true together? Why can't subcontraries be false together?
7. Evaluate the following as a possible definition for contradiction: the relation between propositions with the same subject and predicate which differ in both quantity and quality.

8. How does subalternation differ radically from any mode of opposition?
9. Why must a true subalternand necessitate a true subalternate, but not vice versa? And why must a false subalternate necessitate a false subalternard, but not vice versa?
10. How can the truth of an A proposition necessitate the truth of the corresponding I, since an A seems to demand only the possible existence of its subject while the I seems to express the actual existence of its subject?
11. What is common to the legitimate accidental converse and legitimate simple converse? How do they differ? How does this determine the way in which convertend and converse are mutual indices of the other's truth value?
12. Can *Every man is a rational animal* be simply converted? Discuss.
13. How are obversion and simple conversion alike? How do they differ? What advantage does obversion have over simple conversion? What advantage does simple conversion have over obversion?
14. When can something other than the contradiction of the predicate of the obvertend be used as predicate of the obverse in legitimate obversion?
15. What, in general, is the significance of contraposition? What precise function does contraposition have? Why does the contraponend have exactly the same truth value as the contrapositive? Why can't every proposition be contraposed?
16. Discuss each of the propositional relations taken up so far in the context of the singular proposition.
17. For each of the following propositions, (a) restate the proposition in logical form if this is necessary; (b) determine the extension of its subject and predicate; (c) classify it according to quantity and quality; (d) formulate the several propositions related to it by way of opposition, subalternation, conversion, obversion, and contraposition; and (e) suppose the proposition as true (then as false) and indicate the corresponding truth values of the related propositions:
17.1. All mothers love their children.
17.2. Some liquids are more dense than water.
17.3. Every prudent athlete observes the rules of good health.
17.4. The fireman is an important man in the community.
17.5. Man is sentient.
17.6. Everything worth doing is worth doing well.
17.7. Not every vital principle is an immortal soul.
17.8. Birds fly.
17.9. Some polygon is equiangular.
17.10. No polygon is a circle.
17.11. No king who does not respect his obligations to his people is worthy of his crown.
17.12. Some students who do not study are taking the place in the school rightfully belonging to others not fortunate enough to be in the school.
17.13. Some men are just.

RELATIONS BETWEEN PROPOSITIONS

17.14. All athletes who are not at least six feet tall are not unable to play a good game of basketball.
17.15. Some propositions which are not openly copulative are not occultly copulative.
17.16. All beings-of-the-reason which are not negations are relations.
17.17. Water boils at 100 degrees centigrade at sea level.
17.18. Tom's handwriting is illegible.
17.19. Not every man who speaks well of you to your face can be counted upon as a friend who will support you when things are not going your way.
17.20. Every contradictory of the obverse of an *E* proposition is the subcontrary of an *I* proposition.

18. Indicate the truth value of the second proposition in the light of the truth value of the first by marking T, F, or ? in the parentheses following the second proposition in each pair of propositions. Then explain your answer by stating the relationship the second bears to the first in the space following the parentheses.
    18.1. Some A is B. (T)   Some B is A. ( ) _____
    18.2. No A is non-B. (F)   Some non-B is not A. ( ) _____
    18.3. This A is B. (T)   This A is not B. ( ) _____
    18.4. Not every non-A is B. (F)   Not every non-B is A. ( ) _____
    18.5. This non-A is B. (T)   This non-A is not non-B. ( ) _____
    18.6. Every A is B. (T)   Some A is not B. ( ) _____
    18.7. Some A is non-B. (F)   Not every A is non-B. ( ) _____
    18.8. Every non-A is B. (T)   Some B is non-A. ( ) _____
    18.9. This A is non-B. (T)   Some A is non-B. ( ) _____
    18.10. No A is B. (F)   Not every A is B. ( ) _____
    18.11. All A is B. (T)   All non-B is non-A. ( ) _____
    18.12. Every A is non-B. (F)   Some A is non-B. ( ) _____
    18.13. Not every A is B. (T)   Some A is not B. ( ) _____
    18.14. Some non-A is B. (F)   Every non-A is B. ( ) _____
    18.15. This A is B. (F)   Every A is B. ( ) _____
    18.16. Some A is B. (T)   Some non-B is non-A. ( ) _____
    18.17. Every A is non-B. (F)   No A is non-B. ( ) _____
    18.18. This A is not non-B. (T)   No non-B is this A. ( ) _____
    18.19. Some A is not non-B. (F)   Some non-B is not A. ( ) _____
    18.20. A is B. (T)   B is A. ( ) _____

19. State the proposition called for in each of the following. Then indicate whether it is true, false, or questionable in the light of the original, taking the original once as true and a second time as false.
    19.1. The obverse of the simple converse of the contradictory of the contrary of the contradictory of *Not every A is B*.
    19.2. The superior of the simple converse of the subcontrary of the obverse of *Some propositions are hypothetical*.
    19.3. The contrary of the obverse of the contradictory of the contrapositive of *Some liquid is not inflammable*.
    19.4. The accidental converse of the contradictory of the obverse of the simple converse of the inferior of *Every A is non-B*.

THE LOGIC OF THE SECOND OPERATION

19.5. The superior of the contradiction of the obverse of the contrary of the contradiction of *Some whole number is odd.*

20. Criticize the following:
    20.1. Since every law is reasonable, it follows that it is true that some law is reasonable, that every nonreasonable thing is something which is not a law, and that no law is something which is not reasonable; and it follows that it is false that some law is not reasonable, that some law is nonreasonable, and that everything reasonable is a law.
    20.2. Since it is true that every teacher is not a genius, it is true that some teacher is a genius, and false that no teacher is a genius.
    20.3. The following propositions can all be true together, but they cannot all be false together: Some second intentions are beings-of-the-reason; some beings-of-the-reason are not second intentions; every second intention is a being-of-the-reason; every being-of-the-reason is not a second intention.
    20.4. It is not true that all Hibernians are not drinkers, because it is certainly true that several are.
    20.5. Since it is true that some logic teachers are not idiots, it follows that some are idiots. But if some are idiots, it must be false that none of them are idiots. But this can't be false unless it is true that they are all idiots. Hence, we can establish the fact that all logic teachers are idiots from the fact that some logic teachers are not idiots.
    20.6. Some noncombatants are nurses follows from the fact that no nurses are combatants. For if it is true that no nurses are combatants, it is false that some nurses are combatants, and false that some nurses are not noncombatants, and true that some nurses are noncombatants, and finally true that some noncombatants are nurses.
    20.7. Some husbands have wives, if all wives have husbands.
    20.8. The obverse of the simple converse of the inferior of the contrary of a true *E* proposition is a false *O* proposition.
    20.9. The inferior of the contrary of the contradictory of the simple converse of an *I* proposition is merely the simple converse of that *I* proposition.
    20.10. If it is true that some animal is not man, it is false that every animal is man, and it is true that some nonanimal is not nonman, and it is false that no animal is nonman, and it is false that some animal is not nonman, and, finally, it is false that some man is not animal.

PART III

The Logic of the Third Operation

CHAPTER XIV

Reasoning: The Third Operation of the Intellect

I. The Nature of Reasoning

The intellect is made for truth. This, as we have seen, is grasped in the second operation of the intellect, in judgment. Why then is a third operation required? The reason is that the human intellect cannot know all truth at once. In order to fulfill its potential for truth, the human intellect must have recourse to a third operation which, paradoxically, presupposes the second while being for the sake of the second. To understand this, let us recall a few of the observations which were made earlier with respect to judgment. There we saw that a judgment is true only if motivated by something extramental which demands intellectual assent. This motive is called evidence. We saw also that one type of evidence is immediate evidence and that immediate evidence is divided into the self-evident and the factually evident, with an example of a self-evident truth being *The whole is greater than any of its parts* and an example of a factually evident truth being *This typewriter is gray*. It is, however, obvious that we know many things which are neither self-evident nor factually evident. Consequently, in order to assent to the truth of these things a motive other than immediate evidence is required. One possible motive for such assent is faith, which is based on evidence extrinsic to the truth in question. But we are not concerned here with such extrinsic evidence as faith, however important it is in human life. Hence we must search for another motive for assent, and we find this in evidence intrinsic to the truth in question, but not immediately evident to our intellect, either by reason of being self-evident or factually evident. Some propositions cannot be immediately seen to be true, but they can be seen to follow as true from other propositions which are seen immediately to be true. To show how such propositions do follow from propositions immediately evident is to reason. The operation of reasoning, then, is a movement of the mind to a new truth seen in the light of older truths. For example, consider the proposition *Every man is subject to death*. As it stands this is neither self-evident nor factually evident. However, it is self-evident that every composite being is subject to death (dissolution) and that every man is a composite being, and it is impossible to consider these two truths together without seeing in them the truth of *Every man is subject to death*. As a matter of fact, *Every man is subject to death* follows not only from the two self-evident propositions given above; it follows also from the fact that this man is subject to death (which is immediately evident when he is observed to die) and this other man is subject to death, and this other man, and so on until a sufficient enumeration of singulars has been achieved so that it *becomes evident* that every man is subject to death. Here is a proposition, then, which is neither self-evident nor factually evident. In itself, it is not immediately evident, but it can be made evident in the light of both self-evident and factually evident propositions.

Thus, if a man is to realize his full potential intellectually, he must sometimes move from the second operation of the intellect to the third. Yet he does so only so that he might find the evidence which moves him to assent to a proposition. The third operation presupposes the second; yet it is ordered to the second. Judgment remains the intellectual operation par excellence. The necessity to reason is a sign of the intrinsic weakness of the human intellect.

The conclusion in a reasoning process is said to follow from its premises. Thus it depends on the premises. However, its truth is not the same truth as any one of its premises. It is a new truth which exists potentially in its premises, and it is known by anyone who sees how the premises manifest the conclusion. It is potentially in the premises and it is caused by them. Thus, when we speak of a conclusion following from its premises, we do not mean either that it is merely a novel statement of an old truth (i.e., that it was in the premises actually all the time), or that it is simply thought after something else is thought (i.e., that perhaps it was in the premises not at all). The third operation of the intellect is properly understood only as a true movement of the mind from propositions already known to a new proposition potentially contained in them.

II. The Argument and Its Elements

Mental and verbal arguments are related to reasoning in the same way as mental and verbal terms are related to simple apprehension and mental and verbal propositions to judgment. The argument is the composite expression which signifies the object of the third operation of the intellect. Just as the categorical proposition is resolved into terms joined by a copula, so the argument is resolved into propositions related to one another in a certain way. In the act of reasoning the intellect moves from previous knowledge to new knowledge. That part of the argument which expresses the previous knowledge is called the antecedent. The propositions making up the antecedent are called premises. That part of the argument signifying the new knowledge is the consequent or conclusion. We have used the terms "antecedent" and "consequent" before in a similar but not identical fashion. We spoke of the antecedent and consequent in the conditional proposition. The usage is the same in this sense: neither antecedent can be true without its consequent being true. The difference is clear, however. The antecedent in the argument is composed of more than one proposition which together can be seen to necessitate the consequent. The argument is more than one proposition, even one compound proposition; it is a composite expression made up of several propositions which retain their identity as distinct propositions and which together compose a logical entity which is different from any conditional proposition. The conditional proposition is an assertion that of two component propositions the first cannot be true without the second being true. The first of these two component parts is the antecedent of the conditional, and, unlike the antecedent of the argument, it need not be sufficient to yield the consequent as a conclusion. It should be clear, then, that the terms "antecedent" and "consequent" have their own distinctive meaning when they are taken determinately in reference to the argument. The antecedent of the argument is composed of the propositions which serve as premises and, accordingly, it expresses the previous knowledge from which the conclusion is generated. The consequent of the argument expresses the conclusion which follows from the antecedent.

Reasoning is the third operation of the intellect, which presupposes the second, which in turn presupposes the first. Similarly the argument is resolved into propositions and these in turn into terms. The proximate matter of the argument is the number of propositions from which it is constructed; its remote matter is the number of terms out of which the propositions are ultimately constructed.

Any movement of the reason from previous knowledge to new knowledge is, in a general sense, discursive. Thus we previously spoke of both definition and division as instances of discourse in the sense that they bring the mind from a less perfect to a more perfect grasp of a concept.

The most perfect instance of discourse, however, is the argument, which yields a conclusion which was virtually in the premises but is actually different from either of them. Definition and division result in a new way of knowing an old object, whereas argument yields an entirely new object for the intellect.

III. Validity and Truth

In an argument the conclusion follows from the premises. This means that the premises are so interrelated that they manifest the conclusion, and that the premises cannot be true without the conclusion being true. The validity of an argument rests on the connection between its antecedent and consequent. It follows that validity is a strict property of argumentation. There is no argument which is not valid, because what essentially constitutes an argument necessarily makes it valid. If there is no validity, i.e., if the conclusion does not follow from the premises, there simply are no premises and conclusion, no antecedent and consequent. Strictly speaking, from the point of view of validity (which is the concern of formal logic) there are no bad arguments, no arguments in which a conclusion does not follow from the premises. Lacking validity, a composite expression is not an argument at all, and it has neither conclusion nor premises. Nevertheless, it is customary to speak of invalid argumentations and to note that in an invalid argumentation the conclusion does not follow from its premises. This mode of speaking is common enough so that we have adopted it here. Nevertheless, it is acceptable to speak in this fashion only if a more strict usage of the terms is recognized. To speak of an invalid argumentation is legitimate only so long as it is understood that we speak of a composite expression which is an *attempted* argument which falls short of being genuine because the propositions proposed as premises do not in fact yield the proposition proposed as a conclusion.

Validity and truth do not necessarily go together. An argument can be valid even though none (or only some) of its propositions are true. An attempted argument can be invalid even if each of its propositions is true. Yet there is one connection between validity and truth which is extremely important. In fact, if it were not for this connection there would be no reason for perpetuating an interest in logic, or in any science for that matter. The connection is this: given a valid argument, a true antecedent necessitates a true consequent. The relation between premises and conclusion is that of cause and effect. As long as a cause exists (and exists as a cause), its effect exists. Since premises and conclusions are propositions, the being of each is a being-true. Thus, so long as the premises are true (and so long as they function as premises), the conclusion must be true. If this were not the case, that is, if one could begin with true premises and, arguing validly, still reach a false conclusion, there would be no purpose in reasoning, for one could never be sure of his conclusion.

A true antecedent necessitates a true consequent. However, a false antecedent does not necessitate a false consequent. Given an antecedent with one (or all) of its propositions false, the conclusion which validly follows from it may or may not be false. It is quite possible for it to be true. Consider the following valid argumentations:

1. Every three-sided plane figure is rational.
   But every brute is a three-sided plane figure.
   Therefore, every brute is rational.
2. Every three-sided plane figure is rational.
   But every man is a three-sided plane figure.
   Therefore, every man is rational.

Neither antecedent is true, while the consequent of the first is false and the consequent of the second is true. Clearly a false antecedent can generate a conclusion which is either true or false. Since this is the case, we cannot say that false premises cause a conclusion to be either false or true. Some conclusions from false premises happen to be false, and some happen to be true. We can only say that false premises, as premises, cause a conclusion which would be true if these premises were true. But as false, they can cause nothing, since to be false is, for a proposition, not to be. The truth value of a conclusion following from false premises will have nothing at all to do with the truth value of the premises.

Because a true antecedent in a valid argumentation necessarily yields a true conclusion, it follows that a false conclusion can have come only from a false antecedent. And because a false antecedent might happen to generate a true conclusion, it is impossible to determine the truth-status of an antecedent simply from the fact that its consequent is true. The true consequent might have come from either a true or false antecedent. Students are cautioned especially to remember that a true consequent does not imply a true antecedent. It is quite possible to accept a conclusion as true (e.g., the Communist conclusion that any exploitation of the workingman must cease) without accepting its premises as true (namely, the Marxist-Leninist theory of economic evolution). That day and night will succeed one another follows from the Ptolemaic theory on astronomy, and also from the Copernican theory, which contradicts that of Ptolemy. If one begins with premises which are known to be true, he can be sure of a true conclusion. When one begins with knowledge about the truth-status of a conclusion he can only say that if the conclusion is false the premises themselves must be false.

IV. The Division of Argument

The meaning of argument should become clearer if we consider its types. Besides, the several divisions will help to determine the content and order of succeeding chapters in this part of our textbook.

Induction and Deduction

One of the most significant divisions of argument is into inductive and deductive. Reasoning, as we have said, shows how a mediately evident proposition is related to those immediately evident propositions which manifest its truth. Generally speaking, resolution into self-evident propositions is deductive, and resolution into factually evident propositions is inductive. There are, however, many exceptions to this. Only the perfect form of deduction, which we shall call demonstration, employs self-evident premises. Factually evident propositions can be employed as premises in deduction as well as induction. As a matter of fact, both deduction and induction allow for propositions which are not certainly evident at all, but which are only probable. Indeed, from the point of view of formal logic, deduction and induction can be discussed independently of the truth-status of the propositions used in argument. Nevertheless, the general point stands. Deductive argument proceeds from the more universal to the less universal. A deductive conclusion is a less universal proposition which can be resolved into the more universal premises which virtually contain it. And the self-evident proposition is the prime example of the universal premise. However, from the formal point of view, all that is required for deductive argument is a movement purely on the intelligible plane from a universal premise to a less universal conclusion. The following argument is formally deductive:

>   Every A is B.
>   But every C is A.
>   Therefore, every C is B.

Every C is B is a less universal proposition which falls under *Every A is B*. The reason why it does is that C is a subjective part of A. Thus, when *Every A is B* is coupled with *Every C is A*, it becomes evident that

Every C is B. This is a deductive movement of the mind. The argument is called deduction or syllogism. Induction, on the other hand, proceeds to the universal from the singulars of sense experience which represent concretizations of the universal. Given a sufficient enumeration of singulars, the intellect can move from the sensible plane to the universal plane and form a proposition *universally* affirming the predicate found in each of the singulars of the whole of which the singulars are subjective parts. An inductive conclusion is a universal proposition which can be resolved into a number of singular propositions sufficiently enumerated to manifest the universal conclusion. Thus, *Every A is B* might be resolved back into *This first A is B,* and *This second A is B,* and *This third A is B,* etc., etc.

Categorical and Hypothetical Argument

We know that arguments are composed of propositions. And we know that some propositions are categorical and others hypothetical. Some arguments are categorical because each of the propositions involved is categorical. Others are hypothetical because the principal premise at least is hypothetical. The first example below is a categorical syllogism. The second is a hypothetical syllogism.

1. Every A is B.
   But every C is A.
   Therefore, every C is B.
2. $a \rightarrow b$
   But a
   $\therefore, b$

Demonstrative, Dialectical, and Sophistical Argument

Some arguments yield a certainly true conclusion. These are called demonstrations. Others give a probably true conclusion. These are dialectical arguments. Others give only an apparently worthwhile conclusion. These are fallacious or sophistical. Strictly speaking, we should divide and then subdivide in order to make this a meaningful division. Arguments are either sound or defective. Defective arguments are, generally speaking, fallacious or sophistical. This deficiency may be on the side of form, or matter, or both. On the side of form, sound arguments must be valid. On the side of matter, sound arguments will differ in so far as one proceeds validly from absolutely certain premises to a conclusion which cannot not-be true, while the other proceeds validly from premises which can, at best, establish only a probable conclusion. The first type of sound argument is demonstrative; the second, dialectical. Demonstration, of course, is more perfect than dialectic. However, dialectical argument is worthwhile. It is, in fact, highly significant, since in many areas the human intellect is unable to reach demonstration and must be satisfied with dialectic. Note that the difference between demonstration and dialectic is material rather than formal. The first example below is demonstrative; the second, dialectical; and the third, sophistical or fallacious:

1. Every rational animal is capable of speech.
   But every man is a rational animal.
   Therefore, every man is capable of speech.
2. Every mother loves her child.
   But Marguerite is a mother.
   Therefore, Marguerite loves her child.
3. Marquette students number over 10,000.
   But John is a Marquette student.
   Therefore, John numbers over 10,000.

Immediate and Mediate Inference

Some logicians distinguish between immediate and mediate arguments. This is a false division since there are no immediate arguments or inferences. The conclusion in an argument never comes from one premise alone. There are always at least two premises. The conclusion comes from the principal premise, but only by way of another premise seen under the principal premise. Alleged cases of inferences from one proposition to another are really instances of restating a truth in another way. But this is not reasoning. For example, to move from an *I* proposition to its simple converse is a legitimate and immediate logical movement. However, the converse of the *I* is not a conclusion from the *I*. It is only the original truth expressed in a new proposition. As far as argumentation is concerned there are only mediate argumentations.

Exercise XIV

1. Define: reasoning (or inference), discourse, immediate proposition, mediate proposition, argumentation, antecedent, consequent, validity, truth, deduction(or syllogism), induction, categorical argumentation, hypothetical argumentation, demonstrative argumentation, dialectical argumentation, sophistical or fallacious argumentation.
2. Explain the reason why the third operation of the intellect is necessary for intellectual perfection.
3. Explain the dependence of the third operation upon the second and its simultaneous orientation to the second.

4. In what sense can the acts of defining and dividing be spoken of, along with the act of reasoning, as discursive?
5. In what sense is validity independent of truth, and in what sense are they necessarily interrelated?
6. Supply examples of valid argumentations with true antecedent and true consequent, false antecedent and false consequent, and false antecedent and true consequent.
7. Evaluate the several divisions of the argumentation suggested in this chapter by reference to the rules for a good division. Are these divisions physical or logical? What is the precise principle of division in each case?

CHAPTER XV

The Nature of the Categorical Syllogism

I. The Definition of the Categorical Syllogism

In an argument a consequent follows from an antecedent made up of a given set of related propositions. The propositions in the antecedent are premises. The consequent is the conclusion. An argument with an antecedent which is more universal than the conclusion is deductive. It is called a syllogism. A syllogism which shows the connection between the predicate and the subject of its conclusion by relating each of them to a common third term in its antecedent is a categorical syllogism.

Consider the following proposition: *Every man is a responsible agent.* This proposition is neither factually evident nor self-evident. Yet the mind can be moved to assent to the truth of this proposition if its subject-term (*Every man*) and its predicate-term (*responsible agent*) are joined to a common third term in antecedent propositions or premises. Such antecedent propositions in the present instance could be: *Every intelligent being is a responsible agent* and *Every man is an intelligent being.* In these propositions the subject- and predicate-terms of our original proposition are joined to a common third term, in this instance *intelligent being.* Expressing our premises and conclusion symbolically, we have: *Every A is B, Every C is A,* and *Every C is B.* Here we see that B belongs universally to C because we have previously seen that B belongs universally to A and that A belongs universally to C. By linking the terms of our original proposition (*Every man* [C] and *responsible agent* [B]) to a common third term (*intelligent being* [A]) in two previous propositions, we see that it must be true. If one of the terms of our conclusion had been identified with the third term and if the other term had been separated from it in the premises, then we would have shown that those terms must be separated from each other in the conclusion.

The distinctive feature of the categorical syllogism is its third term. The antecedent shows that the predicate of the conclusion must either be affirmed or denied of its subject in the light of the way in which each is logically related to the same third term. A categorical syllogism is defined as a deductive argument which shows how the terms of its conclusion are related to one another by showing how each is related to a common third term in its antecedent.

II. The Elements of the Categorical Syllogism

A categorical syllogism is so called because each of the three propositions of which it is composed is categorical. Two of these three propositions are contained in the antecedent of the syllogism as premises. The third is the consequent or conclusion of the syllogism. The subject-term and the predicate-term of the conclusion are known as the extremes of the syllogism. Ordinarily the predicate-term of the conclusion is called the major extreme or major term, and the subject-term the minor extreme or minor term. The third term, which does not appear in the conclusion but which is found in each of the premises, is called the middle term. The major term is related, either as subject or predicate, to the middle term in what is called the major premise. The minor term is related, either as subject or predicate, to the middle term in the minor premise. The middle term is so called because it functions as the principle *through which* the extremes are seen to be related to each other either by way of affirmation or denial. The major and minor terms are given these names because the major is related to the minor as its logical superior; at least it is normally apt to be affirmed or denied of the minor term. The premises of a categorical syllogism are so related that one, the major, is more universal than the other, the minor, and both shed light on a conclusion which is less universal than the major premise. The major term is related to the middle term in the major premise and the minor term is related to the middle in the subordinate or minor premise. The names for these terms are especially appropriate if we look at them in the most perfect form of categorical syllogism. Here, as we shall see, the major term is related to the others as the most extended of the three, the minor as the least extended of the three, and the middle as less extended than the major but more extended than the minor.

It is best to express the major premise first and the minor premise second. This helps show that the minor premise should be taken as less extended than the major premise and as subordinated to it. This point can be further emphasized by introducing the minor premise with some word such as the conjunction "but." The conclusion is ordinarily last and introduced by an adverb such as "therefore" to indicate its status as a consequent. It is not necessary that categorical syllogisms be expressed grammatically in precisely this fashion, and one should be prepared to encounter them expressed in a number of different ways. However it is ordinarily advisable, especially for students in elementary logic, to restate syllogisms according to the plan just suggested when they are encountered in some other form. As long as the conclusion of a syllogism can be discerned, it is easy to restate the syllogism grammatically in order to express its true logical pattern as clearly as possible. The major and minor terms are clear from the conclusion, and from them one can easily locate the middle term and the major and minor premises. This is all one need know in order to state the syllogism in strictly logical form.

The following example can be used to point out the elements common to every syllogism. Because it is an example of the first figure it can be seen most perfectly to illustrate the reasons why the major, minor, and middle terms have the names they have. It also serves to illustrate how the minor premise is subordinated to the major, in conjunction with which it yields a conclusion less universal than the major premise.

```
                   ⎧  Every animal is an organism. — Major premise
                   ⎪         M (middle         T
    Antecedent     ⎨           term)
                   ⎪
                   ⎩  But every man is an animal. — Minor premise
                              t                M

Consequent — Therefore, every man is an organism. — Conclusion
                              t (minor    T (major
                                term)      term)
```

III. Direct and Indirect Conclusions

The major term of the categorical syllogism usually appears as the predicate in the conclusion, where it is either affirmed or denied of the minor term. The major term is normally used in this way because it is normally suited or apt to be used in this way. Consider *organism* and *man*. Organism is a remote genus of *man*; as such, it is calculated to be said of *man* rather than vice versa. Thus we say *Every man is an organism*. However, we can legitimately convert this to *Some organism is a man*. Here the predicate is not a term naturally suited to be a predicate for the subject to which it is related. Nevertheless, it can be used as the predicate so long as the subject is taken particularly. *Every man is an organism* is called a *direct* proposition, because the term used as the predicate is naturally suited for this function. On the other hand, *Some organism is a man* is an *indirect* proposition because the term normally suited to be used as the predicate is taken as the subject. Both of these propositions follow as consequents from an antecedent including the propositions *Every animal is an organism* and *Every man is an animal*. *Every man is an organism* is the more natural conclusion. We can call it a *direct* conclusion. *Some organism is a man* is a less natural conclusion. However, it is a legitimate *indirect* conclusion. A direct conclusion differs from an indirect conclusion by using the major term as the predicate, as it would ordinarily be used, whereas an indirect conclusion uses the major term as the subject.

We shall see that there are three basically different patterns of syllogistic procedure. These are called the figures of the categorical syllogism. Both direct and indirect conclusions are possible in each of the three figures.

IV. The Figures of the Categorical Syllogism

We have spoken of the perfect form of the categorical syllogism, while allowing for the possibility of less perfect forms. What are its various forms, and why is one more perfect than the others? There are three basically different forms, called figures, of the categorical syllogism. Each of these admits of variations of form within itself, known as the moods of the syllogism. We shall speak first of the figures and then of the moods.

The most important single principle of syllogistic procedure is the middle term through which the major is united to or separated from the minor term. The reason why there are three basically different patterns or figures of syllogistic procedure is this: there are three, and only three, basically different ways of relating the middle term to the major and minor terms. The major term is always taken as more extended than the minor. Understanding the major term in this way there are three basically different ways in which the middle term can be related to the extremes. The middle can be logically less extended than the major and more extended than the minor, or more extended than either the major and minor terms, or less extended than either the major or the minor terms. These three possible relations of the middle term to the extremes give us, respectively, the first, second, and third figures of the categorical syllogism. There is no fourth figure in which the middle term would be more extended than the major and less extended than the minor. This is an impossible combination, since the major is by definition taken as more extended than the minor. A fourth figure would demand the impossible, namely, one term both less and more extended than another term or, in other words, a major term which is not a major term. There is another way of expressing the three different ways in which the middle term can be related to the extremes. In the first figure the middle is affirmed or denied of the minor while the major is affirmed or denied of the middle. In the second figure the middle is affirmed or denied of both the major and minor. In the third figure both the major and the minor are affirmed or denied of the middle. The syllogistic patterns for each of the three figures can be diagramed as follows:

	(1)	(2)	(3)
Major premise	M — T	T — M	M — T
Minor premise	t — M	t — M	M — t
Direct conclusion	t — T	t — T	t — T
Indirect conclusion	T — t	T — t	T — t

The following examples illustrate each of the three figures, with both direct and indirect conclusions noted:

First figure — direct conclusion
Every rectangle is four-sided.
    M   T
But every square is a rectangle.
       t   M
Therefore, every square is four-sided.
          t   T

First figure — indirect conclusion
Every rectangle is four-sided.
    M   T
But every square is a rectangle.
       t   M
Therefore, some four-sided thing is a square.
           T   t

Second figure — direct conclusion
No plant is sentient.
   T   M
But every animal is sentient.
     t   M
Therefore, no animal is a plant.
       t   T

Second figure — indirect conclusion
No plant is sentient.
   T   M

But every animal is sentient.
       t   M

Therefore, no plant is an animal.
            T   t

Third figure — direct conclusion

Every good scientist has a sincere respect for truth.
      M   T

But some good scientist is American.
         M   t

Therefore, some American has a sincere respect for truth.
            t   T

Third figure — indirect conclusion

Every good scientist has a sincere respect for truth.
      M   T

But some good scientist is American.
         M   t

Therefore, someone who has a sincere respect for truth is
                                            T
American.
   t

The first figure is said to be *first* because it is the most perfect form of the categorical syllogism. It is the form of syllogistic procedure most natural to the human reason. This is evident from experience. It can also be seen in the fact that the basic principles of syllogistic procedure can be applied directly only to the first figure. We shall say more of this shortly. The first figure, as we have seen, is the only figure in which the middle term is a middle in every respect, falling between the extremes as far as extension is concerned. It is for this reason the only figure in which the major is the most extended term while the minor is the least extended. We shall see that the first figure is also the only figure in which a universal affirmative conclusion is possible. For many reasons, therefore, it is the most perfect form of categorical syllogism. The second figure is put before the third because in it the middle term is *more* extended than the other terms, whereas the middle is the *least* extended of the terms in the third.

Some logicians hold that there is a fourth figure of the categorical syllogism in which the middle term is the predicate of the major premise and subject of the minor premise. This argument would hold if the figures of the categorical syllogism were determined according to the different ways in which the terms of a syllogism could be grammatically arranged. There are four different ways *grammatically* to express the categorical syllogism. But since there are only three ways in which the middle term can be *logically* related to the other two terms, there can be only three different logical figures. The fourth way of grammatically expressing the syllogism ordinarily has the force of the first figure concluding indirectly. Understood as such, it is an acceptable grammatical expression of a legitimate logical procedure. It would be a more faithful expression of this logical procedure, however, to transpose the premises, putting the major premise first with the minor premise expressed along with and under the major premise. For example:

Every man is an animal.   Every animal is an organism.
Every animal is an organism.   Every man is an animal.
Some organism is a man. ⟶ Some organism is a man.

V. The Moods of the Categorical Syllogism

The figures of the categorical syllogism depend upon the disposition of the *terms* in the premises. The moods of the categorical syllogism depend upon the disposition of the *premises* according to quantity and quality. Each syllogism involves two premises. Each premise is either universal affirmative (A), universal negative (E), particular affirmative (I), or particular negative (O). Singular premises (A^1 or E^1) do occur as minor premises in the first and second figures. However, they present no great difficulty and do not constitute any distinct mood. For all practical purposes they are handled as though they were universal propositions. Disregarding the singular premises, we see that there are sixteen possible moods within each figure. The first figure can have a major premise which is an A proposition coupled with an A, E, I, or O minor. It allows also for an E, I, or O major as well, each with the same possibilities for the minor premise. The same is true for the second and third figures. The moods of the categorical syllogism are identified ordinarily by describing them in terms of A, E, I, and O. The description for a valid mood usually includes the conclusion as well as the premises of the syllogism. For example:

1. (A) Every intellectual virtue is a quality.
                  M   T
   (A) But every science is an intellectual virtue.
              t   M
   (I) Therefore, some quality is a science.
                 T   t

This is a first figure syllogism concluding indirectly with the mood AAI.

2. (E) Nothing which fits into a category is a being-of-the-reason.
        T   M
 (I) But some relation is a being-of-the-reason.
      t   M
 (O) Therefore, some relation does not fit into a category.
         t   T

This is a second figure syllogism concluding directly with the mood EIO.

3. (I) Some triangle is isosceles.
     M   T
 (A) But every triangle is a figure possessed of three angles.
       M   t
 (I) Therefore, some figure possessed of three angles is isosceles.
           t   T

This is a third figure syllogism concluding directly with the mood IAI.

Not every mood is valid. Certain moods are not valid in any of the figures and most are invalid in one or another figure. Once we have determined the rules for the validity of the categorical syllogism we shall measure each of the possible moods against these rules and discover that only a small fraction of them can yield a valid conclusion in any figure.

VI. The Basic Principles for Categorical Syllogism

An affirmative conclusion in a valid categorical syllogism follows from its premises in virtue of the general principle of *triple identity*: *two things identified with the same third thing are identified with one another*. A negative conclusion follows in virtue of the principle of the *separating third*: *two things one of which is identified with a third thing while the other is separated from it are separated from one another*. The principles of triple identity and of the separating third cannot be called into doubt, for their denials are patent absurdities. Syllogistic procedure is sound precisely as grounded in these principles. Any syllogism is valid only so long as it measures up to the requirements of either of these principles. The principles are expressed in terms of comprehension. They are best applied to the categorical syllogism by way of two principles expressed in terms of extension which are more particularly proportioned to the syllogism. The first of these is known as the *dictum de omni* (literally, "what is said of all"): *whatever is universally said of a logical whole must be said of each of its subjective parts*. The second is the *dictum de nullo* ("what is said of none"): *whatever is universally denied of a logical whole must be denied of each of its subjective parts*.

The *dictum de omni* and the *dictum de nullo* follow from the principles of triple identity and of the separating third respectively. Whatever is said of a logical whole must be said of its subjective parts because the whole itself is said of each of these parts. Consequently, something which is identified with a logical whole must be identified with its subjective parts in accord with the demands of the principle of triple identity. Consider the example: *Every animal is an organism; but every man is an animal; therefore, every man is an organism.* The comprehensive notes of *organism* are seen to belong to any subject possessed of the comprehensive notes of *man*, because the comprehensive notes of *organism* belong to any subject possessed of the comprehensive notes of *animal*, which is the case with *man*. The comprehension of *man* and the comprehension of *organism* are each seen to be identical in subject with the comprehension of *animal* and thus identical in subject with one another. This is an evaluation of the syllogism directly in terms of the principle of triple identity. It can just as legitimately, and more easily, be evaluated directly in terms of the *dictum de omni*. Since *organism* is said of any subject of which *animal* is said, it must be said of *man*, which is an *animal*. The example is an instance of a valid categorical syllogism. It is seen to be valid precisely because it measures up to the demands of the principle of triple identity and the *dictum de omni*. We can consider a second example: *No rectangle is curvilinear; but every square is a rectangle; therefore, no square is curvilinear.* This syllogism is constructed in accord with the principles of the separating third and the *dictum de nullo*. Comprehensively considered, *curvilinear* is seen as separated from *square*, because *curvilinear* is separated from *rectangle*, whereas *rectangle* and *square* are identified. From the point of view of extension, *curvilinear* is denied of *square* precisely because it is universally denied of *rectangle*, of which *square* is a subjective part.

The two examples of categorical syllogism we have used to illustrate the force of the basic principles of the syllogism are both in the first figure concluding directly. This is significant, for the basic principles behind the syllogism are immediately applicable only to syllogisms in the first figure concluding directly. This does not mean that second and third figure syllogisms can never be valid, but it does mean that they are valid only in so far as they can be reduced to syllogisms in the first figure. The principles of triple identity and the separating third, as well as the *dictum de omni* and the *dictum de nullo*, remain the ultimate canons for the validity of any categorical syllogism. Each of the valid moods in the second and third figures can be reduced to a valid mood in the first figure. The second or third figure mood is seen to be valid because the mood of the first figure to which it is reduced measures up to the demands of these basic principles. Hence, one method for evaluating second and third figure syllogisms is to reduce the syllogism to a mood of the first figure and then to measure the syllogism against the basic principles. We shall soon discuss the reduction of second and third figure syllogisms to syllogisms in the first figure. We shall see that the reduction is not always easy and that it always requires care and precision. Fortunately, there is a second method for evaluating second and third figure syllogisms. Although the basic principles of the categorical syllogism are directly applicable only to first figure syllogisms, we can deduce from these principles a particular set of rules which can be applied to any categorical syllogism in any figure. With this set of rules we need not reduce a second or third figure syllogism to the first figure in order to test its validity. We can do this simply by checking it against the particular rules for the categorical syllogism. Since these are deduced from the basic principles, they represent a true measure of the worth of the syllogism in question. And, in so far as the basic principles remain directly applicable only to the first figure, any valid mood in the second or third figures is valid precisely because it is logically reducible to a valid mood in the first figure. This is true even though we test it against the particular rules without bothering first to reduce it to the first figure.

VII. An Objection to the Categorical Syllogism

The *dictum de omni* requires that whatever is said universally of a given logical whole be said of each of its subjective parts. Thus the categorical syllogism in its affirmative form expresses a movement of the intellect from a universal proposition in which a predicate is said of a given universal subject to a conclusion in which this predicate is said of a less universal subject related to the more universal subject as a subjective part. The conclusion is reached because it is contained in the more universal proposition from which it is generated. It is precisely at this point that the categorical syllogism seems vulnerable to criticism. If the conclusion of a categorical syllogism is already contained in its major premise, then a syllogism does not seem to involve a movement of the intellect. A categorical syllogism would seem to be no more an instance of reasoning than the passage from the *A* or the *E* to the corresponding *I* or *O* by way of subalternation. Categorical syllogism would seem to be guilty of the fallacy of begging the question, that is, of presupposing the conclusion that it presumes to establish.

John Stuart Mill is perhaps the best known author of the charge that every syllogism begs the question.[1] He suggests the syllogism: *All men are mortal; but Socrates is a man; therefore, Socrates is mortal.* He argues that no one can reasonably propose that all men are mortal unless he already knows that Socrates is mortal. Thus, it would seem, the syllogism begs the question by presupposing its conclusion in its major premise. Certainly if Socrates is a man, and if Socrates is not mortal, then it is not true that all men are mortal. We can agree with Mill this far: *All men are mortal* cannot be true unless it is also true that Socrates is mortal. In fact it is because of this that we can conclude to the truth of *Socrates is mortal* from the truth of *All men are mortal.* However, it is not necessary to know that Socrates is mortal in order to know that all men are mortal. *All men are mortal* expresses something other than a factual report. *Mortal* is seen to belong to all men not because every man has been observed and counted, but because human nature is sufficiently understood so that it is seen to allow for the possibility of death. *Man*, in abstraction from the individuating characteristics of individual men, including Socrates, is understood to be such that any subject which is a man is by that token also mortal. To be sure, the proposition *All men are mortal* results from an experience of individual men. It is a universal proposition achieved by way of an induction from singular instances of man and mortality. However, it is a proposition which, in itself, is independent of any singular instance of which it is a universal statement. It does not depend upon the fact that Socrates is mortal. It is more than a report that all men are in fact mortal because each has been observed to be. Thus *All men are mortal* does not presuppose the proposition *Socrates is mortal.* Further, *All men are mortal* in conjunction with the proposition *Socrates is a man* yields the conclusion *Socrates is mortal*. This is a legitimate conclusion. It does not beg the question. *All men are mortal* is true independently of the fact that Socrates is mortal and would be true even if Socrates did not exist. It would not, of course, be true if Socrates were, per *impossible*, not mortal. However, we say *per impossible* here precisely because *All men are mortal* is true quite independently of Socrates, so that if Socrates does exist as a man it is necessary that Socrates be mortal.

Not every proposition with the grammatical form of "All men are mortal" can be a major premise in a categorical syllogism. Some represent simply a factual report on an exhaustive examination of possible subjects. Take the proposition *All the students in the class passed the most recent exam.* This proposition can be made only if the examination grade of

---

[1] See Mill's *System of Logic*, Bk. II, Chap. III (New York: Longmans, Green & Co., 1930).

each student in the class is observed and seen to be passing. This proposition is simply a report on the fact that this student, and that student, and every other student in the class had a passing grade on the most recent exam. There is no necessary connection between being a member of this class and passing the most recent exam. It must be known that each member of the class passed the exam before it can be known that the entire group passed the exam. One could not *conclude* that John passed the exam because John is a member of the class. The fact that John passed the exam is presupposed to the proposed major premise and is, in fact, partially what this proposition explicitly reports.

All men are mortal is independent of *Socrates is mortal*. It can be called a universal proposition involving *true* universality. *All the students in this class passed the most recent exam* is not independent of *John* (i.e., *this student in this class*) *passed the most recent exam*. At best we can call this a universal proposition which is *enumeratively* universal. The middle term in a syllogism, however, functions as such only in so far as it is a true universal, not an enumerative universal. The major, minor, and middle terms may be one in subject (i.e., existentially identified in the same thing), yet they must be different as objective concepts (if not in themselves, at least as conceived) or else the identification of both the major and the minor in the middle causes no advance in knowledge. The minor term in a syllogism whose middle term is only enumeratively universal is not significantly different from the middle term itself. The latter is simply the sum of the individuals of a given class of which the minor is an instance. Thus the major cannot be *communicated* to the minor by way of the middle. Inasmuch as the major term is said of the middle it is already said of the minor.

VIII. The Expository "Syllogism"

The categorical syllogism generates a conclusion by linking the major term to the minor term by way of the middle. The middle term is a principle of communication for the major in reference to the minor. As such, the middle must be a universal term so that it is impossible for a singular term ever to function as a middle term in a categorical syllogism. What, then, do we say of something like this?

>Peter is a Canadian.
>But Peter is a student in my class.
>Therefore, some student in my class is a Canadian.

This seems to be a syllogism, and the first and second propositions cannot be true without the third being true as well. However, Peter does not function as a middle term. There is nothing in Peter, as Peter, which will allow us to relate *student in my class* to *Canadian*. Peter serves rather as a concrete example of someone who is both a student in my class and a Canadian. I point to Peter as a singular example which establishes the proposition *Some student in my class is a Canadian*. The function of Peter is simply to expose the truth that a certain student in my class is a Canadian. Because Peter is not a middle term, this example is not an example of a syllogism. However, because it has the *appearance* of a *syllogism*, and because Peter does serve to expose the truth of what appears to be a conclusion, it is said to be an example of an expository "syllogism."

The middle term in a categorical syllogism cannot be singular if it is to function as a principle of communication. The major term cannot be singular either. The major must be more universal than the minor. However, the minor term can be singular, and in some cases is. It is significant to note, however, that the categorical syllogism is primarily ordered to conclusions concerning universal subjects. It allows for conclusions concerning singular subjects only in so far as the singulars fall under the extension of the universals.

Exercise XV

1. Define: argumentation, syllogism, categorical syllogism, major term, minor term, middle term, major premise, minor premise, direct conclusion, indirect conclusion, figure, mood, true universality, enumerative universality, expository "syllogism."
2. Explain the meaning of categorical syllogism in terms of the elements into which it can be both proximately and ultimately resolved. Define each of these elements.
3. What is the difference between a syllogism concluding directly and a syllogism concluding indirectly?
4. How are the figures of the categorical syllogism determined? Why have some logicians spoken of a fourth figure? What makes a fourth figure a logical absurdity?
5. How many possible moods of the categorical syllogism are there? Explain.
6. What is the evidence for the truth of the basic principles for categorical syllogism?
7. Explain the legitimacy of the following syllogisms in terms of the basic principles for categorical syllogism:
   7.1. Every polygon is rectilinear.
       But every square is a polygon.
       Therefore, every square is rectilinear.
   7.2. No polygon is curvilinear.
       But every square is a polygon.
       Therefore, no square is curvilinear.

8. How are the basic principles for categorical syllogism applied to second and third figure syllogisms?
9. How does the distinction between true universality and enumerative universality help to defend the syllogism against the charge of begging the question?
10. Why is the expository "syllogism" not a syllogism in the strict sense, and of what value is it, logically considered?

CHAPTER XVI

The Rules for the Categorical Syllogism

I. The General Rules for Any Figure

The following rules are particular applications of the basic principles of the syllogism. They are deduced from these basic principles and can be applied to any categorical syllogism in any figure. If any one of them is broken, the attempted syllogism is invalid. An attempted syllogism might violate several of the rules at the same time. However, it is necessary only to observe one infraction to know that the syllogism is invalid. Thus a syllogism must conform to each of the rules in order to be valid.

RULE 1. *There must be three terms and only three terms. These must be so arranged that the middle term is found in one premise with the major term and in the other with the minor term, while the major term and minor term are found in the conclusion.*

This rule follows from the very nature of the categorical syllogism as determined by the basic principles of syllogistic discourse. It simply states the essential pattern of syllogistic procedure.

I cannot argue that some animal is capable of speech because every man is capable of speech and every animal is mortal. Nor can I argue that a certain musical instrument is colorless, odorless, tasteless, and weightless because it is a triangle and because every triangle is colorless, odorless, tasteless, and weightless. The first example obviously employs four terms and is, therefore, not a syllogism. In the second example it *looks* as though there are only three terms, with each used twice, and in the proper arrangement. But "triangle" is used with an analogy so close to equivocity that it does not stand for the same concept each time it is used. Thus the second example also uses four terms. The principle of triple identity demands that each of the three terms in the categorical syllogism be taken generally with the same signification and supposition. Anything less than this constitutes a violation of the very nature of syllogistic procedure just as surely as if four, five, or six completely diverse

THE RULES FOR THE CATEGORICAL SYLLOGISM

concepts expressed in completely diverse ways were employed in constructing the syllogism.

Other examples of syllogisms which are faulty precisely as violating this first rule include:

1. Every triangle is a rectilinear figure.
   But no rectangle is a curvilinear figure.
   Therefore, no triangle is a curvilinear figure.
   (Four terms: [1] *triangle*, [2] *rectilinear figure*, [3] *rectangle*, [4] *curvilinear figure*)

2. Americanism is opposed to Fascism.
   But Fascism is opposed to Communism.
   Therefore, Americanism is opposed to Communism.
   (Four terms: [1] *Americanism*, [2] *opposed to Fascism*, [3] *Fascism*, [4] *opposed to Communism*)

3. Substance is the ultimate genus of animal.
   But every man is substance.
   Therefore, every man is the ultimate genus of animal.
   (Four terms: [1] *substance* taken with logical supposition, [2] *ultimate genus of animal*, [3] *man*, [4] *substance* taken with personal supposition)

4. Marquette students number over 10,000.
   But John is a Marquette student.
   Therefore, John numbers over 10,000.
   (Four terms: [1] *Marquette students* taken collectively, [2] *number over 10,000*, [3] *John*, [4] *Marquette student* taken divisively)

5. Foxes are valuable for their pelts.
   But Rommel was a fox.
   Therefore, Rommel was valuable for his pelt.
   (Four terms: [1] *foxes* taken properly, [2] *valuable for their pelts*, [3] *Rommel*, [4] *fox* taken metaphorically)

6. No A is B.
   But some C is D.
   Therefore, some C is not B.
   (Four terms: [1] A, [2] B, [3] C, [4] D)

7. Every A is non-B.
   But some C is B.
   Therefore, some C is non-A.
   (Five terms: [1] A, [2] non-B, [3] C, [4] B, [5] non-A)

Rule 2. *No term in the conclusion can be universal if it is only particular in the premises.*

This rule follows from the fact that the conclusion of the categorical syllogism is caused by its premises. If any term were used more extensively in the conclusion than it is used in the premises, the supposed effect would be greater than its cause, and this is an absurdity.

I cannot argue that every Midwesterner is an American just because every Badger is an American and every Badger is a Midwesterner. When I attempt to argue in this way, Midwesterner is universal in the conclusion and only particular in the minor premise. This would also be the case were I to argue that no being is necessary because no creature is necessary and every creature is a being. Here again the minor term is more extended in the conclusion than in the minor premise. These syllogisms suffer the fallacy of the *illicit minor term*. To argue that some horses are not four-legged because every horse is four-legged and some animals are not horses violates this same rule. In this case, however, it is the major term which is used universally in the conclusion after being taken only particularly in the major premise. This syllogism involves the fallacy of the *illicit major term*. This is also the case with the syllogism which attempts to show that every substance is a plant because every plant is an organism and every organism is a substance.

Other examples of syllogisms which are defective because of illicit major terms or illicit minor terms include:

1. Every predicable is a being-of-the-reason.
   But some logical relations are not predicables.
   Therefore, some logical relations are not beings-of-the-reason.
   (Illicit major term)

2. No rectangle is a curvilinear figure.
   But every rectangle is a rectilinear figure.
   Therefore, no rectilinear figure is a curvilinear figure.
   (Illicit minor term)

3. Not every substance is an organism.
   But every plant is an organism.
   Therefore, no substance is a plant.
   (Illicit major term — in an indirect conclusion)

4. Every morally responsible agent is a free agent.
   But some morally responsible agents are not men of good will.
   Therefore, some men of good will are not free agents.
   (Illicit major term)

THE RULES FOR THE CATEGORICAL SYLLOGISM   199

   5. Some laws are easily obeyed.
      But no unreasonable ordinance is a law.
      Therefore, some unreasonable ordinances are not easily obeyed.
      (Illicit major term)

   6. No A is B.
      But some C is A.
      Therefore, no C is B.
      (Illicit minor)

   7. Every A is B.
      But every A is C.
      Therefore, every C is B.
      (Illicit minor)

   8. Every A is non-B.
      But some A is not C.
      Therefore, some C is not non-B.
      (Illicit major)

   9. Some A is B.
      But no C is A.
      Therefore, some C is not B.
      (Illicit major)

RULE 3. *The middle term must be universal at least once.*

This rule follows from the principles of triple identity and the separating third. A middle term which is particular both times might stand for one segment of its possible extension the first time and for some other segment of its possible extension the second time. Thus it cannot function properly as a middle term, for it has the effect of being two different terms. The middle term serves its purpose precisely as one term simultaneously related to the predicate-term and subject-term of the conclusion. It is by way of the middle that the predicate-term is communicated to the subject. If the predicate were true only of some instance of the middle while the subject was identified with only some (possibly other) instance of it, then the predicate could not be communicated to the subject by way of the middle. Under these circumstances the middle could not be taken as one term simultaneously related to both. What we have said about a middle intended as a principle of identification holds as well for a middle intended as a principle of separation.

I cannot argue that all men are brutes because all brutes are sentient and all men are sentient, nor can I argue that all men are animals because all animals are substances and all men are substances. Both conclusions are equally invalid, although one is actually true and the other false. To argue that some saints are neurotic because some saints see strange things and neurotics see strange things is to argue invalidly. In each of these instances the conclusion fails to follow from the premises precisely because in none of these syllogisms is the middle term universal at least once. Each can be said to suffer the fallacy of the *unextended middle term*.

Other examples of syllogisms defective because of unextended middle terms include:

1. Every triangle is three-sided.
   But some plane figures are three-sided.
   Therefore, some plane figures are triangles.

2. Some men are intelligent men.
   But some men are blind men.
   Therefore, some blind men are intelligent.

3. Not every substance is quantified.
   But every body is a substance.
   Therefore, not every body is quantified.

4. Every habit is a quality.
   But every virtue is a quality.
   Therefore, every virtue is a habit.

5. Every even number is divisible by two without a remainder.
   But some whole numbers are divisible by two without a remainder.
   Therefore, some whole numbers are even numbers.

6. Some men are mathematicians.
   But some men are Americans.
   Therefore, some Americans are mathematicians.

7. Not every A is B.
   But some A is C.
   Therefore, some C is B.

8. Every A is B.
   But some B is not C.
   Therefore, some C is not A.

9. Every A is non-B.
   But some C is non-B.
   Therefore, some C is A.

Rule 4. *If both premises are affirmative, the conclusion cannot be negative.*

This rule follows immediately from the principle of triple identity, which asserts an *identity* between two things which are *identified* with the same third thing. If one thing could be *separated* from another after each was *identified* with the same third thing, the principle of triple identity could be contradicted. This is impossible.

We cannot argue that no Communists are materialists because all Communists are Marxists and all Marxists are materialists. This conclusion would be absurd in the light of these premises. The contrary of this conclusion follows from these premises. There is little need to multiply examples of this fallacy. To break this rule is so absurd that few are ever tempted to do so and still fewer are fooled by those who do. For this reason examples of this fallacy are relatively trivial.

Rule 5. *Nothing follows from two negative premises.*

This rule follows immediately from the principle of the separating third. According to this principle two things must be separated from one another if one is seen as identified with a third thing while the other is separated from this same third thing. The third thing functions as a principle for separating the other two only if it is positively identified with one and separated from the other. If it were separated from both, there would be no relationship set up between them and the third. In this case it could function neither as a principle of identification nor of separation.

It does not follow that every Communist is a Marxist simply because no Communist can be trusted and no Marxist can be trusted. Nor can we conclude that kerosene is water because neither has a boiling point of 10 degrees centigrade. We might also note that we could not conclude that no Communist is a Marxist in our first example nor that kerosene is not water in our second. When both premises are negative there is no relationship between the extremes and the middle. No conclusion at all is generated, and the supposed premises leave the supposed conclusion decidedly open to either side of a contradiction.

The following examples suffer the fallacy of two negative premises:

1. No triangle is corporeal.
   But some being is not corporeal.
   Therefore, some being is not a triangle.

2. Communists do not favor the continuation of segregation in the South.

But Northern conservatives do not favor the continuation of segregation in the South.
Therefore, Northern conservatives are Communists.

3. No A is B.
   But no C is B.
   Therefore, no C is A.

4. No A is B.
   But not every C is A.
   Therefore, some C is B.

RULE 6. *If one premise is negative, the conclusion must be negative.*

This rule follows immediately from the principle of the separating third. A term positively related to one term and negatively related to another manifests a negative relationship between these two. This rule is the third rule concerned with the quality of the propositions in the syllogism. Together these three rules demand that the two premises be either both affirmative or one affirmative and one negative, and that the conclusion be affirmative in the first case and negative in the second. They rule out the possibility of two negative premises, of a negative conclusion following two affirmative premises, and of an affirmative conclusion when one premise is negative.

In the light of the sixth rule it is impossible to argue that every man is a horse because no horses are endowed with speech and every man is endowed with speech. Once again it is the contrary which follows. And once again it is trivial to multiply examples to illustrate this fallacy. Ordinarily a syllogism which breaks this rule is *obviously* invalid. Thus it is of little value to multiply examples.

These are the six rules which govern the categorical syllogism. If any one of them is violated, the syllogism in question is invalid. Any valid categorical syllogism must measure up to all of them. The first rule merely restates the basic pattern of syllogistic procedure, which lists the elements required in a categorical syllogism and indicates generally the manner in which they must be arranged. The second and third rules set the requirements for the three terms as far as their extension is concerned. The last three rules state the requirements for the propositions used as premises and conclusion from the point of view of quality. No other rules are necessary, but two more can be added to the list. These additional rules are: (7) *Nothing follows from two particular premises,* and (8) *If one premise is particular, the conclusion must be particular.* These are not independent rules. They do not follow directly from the two fundamental principles of the syllogism. They follow indirectly in so far as their violation necessitates the violation of one or more of the six rules already discussed.[1] Thus it is quite possible to evaluate any categorical syllogism simply in the light of the six rules already given. We add a seventh and an eighth to these six for the sake of convenience. Sometimes it is easier to see that the seventh or eighth is violated than to see the accompanying violation of one of the fundamental six. The seventh and eighth rules are legitimate rules, but only in so far as they are involved in the other rules.

Let us consider these additional rules in order to see why they presuppose at least one of the basic six rules.

RULE 7. *Nothing follows from two particular premises.*

If both premises are particular, one of them must be an *I* proposition; otherwise both would be *O* or negative propositions and thus break Rule 5. An *I* proposition, however, has two particular terms. Consequently, the other premise must be an *O* proposition; if it were another *I*, with two particular terms, the middle term would be unextended and violate Rule 3. The conclusion, therefore, must be negative, since one of the premises is negative (Rule 6). But a negative conclusion has a universal predicate. This means that one of the terms in the conclusion would have greater extension than it had in the premises or that the middle term would be unextended (Rules 2 and 3), since the premises, which are *O* and *I*, allow for only one universal term. Thus two particular premises ultimately require either an unextended middle term or an illicit major or minor term.

Each of the following examples violates this seventh rule as well as at least one of the first six rules:

1. Some criminals are psychotic.
   But some Americans are criminals.
   Therefore, some Americans are psychotic.
   (Two particular premises, as well as an unextended middle term)

2. Some plants are green.
   But some organisms are not plants.
   Therefore, some organisms are not green.
   (Two particular premises, as well as an illicit major term)

---

[1] As a matter of fact Rule 4 cannot be violated without the violation of at least one of the other five rules. Nevertheless, Rule 4 is given as a fundamental rule because it follows directly from the principle of triple identity. Each of the other five of the fundamental six rules can be broken independently of the others.

3. Some A is B.
   But some A is not C.
   Therefore, some C is not B.
   (Two particular premises, as well as an unextended middle term and an illicit major term)

Rule 8. *If one premise is particular, the conclusion must be particular.*

A universal conclusion must be either an E or an A proposition. If E, both of its extremes must appear as universal in the premises, since both are universal in the conclusion (Rule 2). Since the middle term must be universal at least once (Rule 3), it follows that only one term can be used particularly in the premises. Moreover, one of the premises must be affirmative (Rule 5) and, as affirmative, must have a particular predicate. According to our supposition, however, one of the premises is particular. Its subject must also be particular. Thus we would have two terms used particularly: one, as subject of the particular premise; the other, as predicate of the affirmative premise. But, as we have noted, there is room for only one particular use of a term in the premises if the conclusion is an E proposition. Consequently an E proposition cannot follow from premises one of which is particular. With an A conclusion the term used as subject of the conclusion is universal and must be used universally in the premise in which it occurs (Rule 2). Since the middle term must be used universally at least once (Rule 3) then there must be at least two universal terms in the premises. Each premise must be affirmative with an A conclusion (Rule 6). A particular premise would have to be an I proposition with both terms particular. The other premise would be an A proposition with a universal subject and a particular predicate. This allows for only one universal use of a term when, as we have seen, at least two are necessary.

Each of the following examples violates this eighth rule as well as at least one of the first six rules:

1. Some good athletes are not big.
   But every good athlete is physically fit.
   Therefore, no one physically fit is big.
   (A universal conclusion with a particular premise, as well as an illicit minor term)

2. Every man is endowed with speech.
   But some two-legged animals are not men.
   Therefore, no two-legged animals are endowed with speech.
   (A universal conclusion with a particular premise, as well as an illicit major and an illicit minor)

3. Every A is B.
But some C is B.
Therefore, every C is A.
(A universal conclusion with a particular premise, as well as an unextended middle term and an illicit minor term)

Let us summarize the important points of this particular section. The basic principles of categorical syllogism, namely, the principles of triple identity and the separating third, and the *dictum de omni* and *dictum de nullo*, are immediately applicable only to syllogisms in the first figure. Syllogisms in the other figures can be checked for validity in two ways: first, by reducing them to the first figure and then measuring them directly by the basic principles; second, by measuring them against a set of rules which are deduced from the basic principles and which are directly applicable to syllogisms in any figure. Here our six fundamental rules come into play, plus two additional rules which are involved in the first six and which can facilitate the evaluation of syllogisms when stated as additional rules. The eight rules are:

1. There must be three terms and only three terms. These must be so arranged that the middle term is found in one premise with the major term and in the other with the minor term, while the major term and minor term are found in the conclusion.
2. No term in the conclusion can be universal, if that same term is only particular in the premises.
3. The middle term must be universal at least once.
4. If both premises are affirmative, the conclusion cannot be negative.
5. Nothing follows from two negative premises.
6. If one premise is negative, the conclusion must be negative.
7. Nothing follows from two particular premises.
8. If one premise is particular, the conclusion must be particular.

II. The Special Rules for Each Figure of the Categorical Syllogism

The eight rules just discussed apply to any categorical syllogism no matter what its figure. Although there is no need to add to them, there are special rules for each figure, and it may be helpful to use them in evaluating syllogisms. These special rules for each figure are not adequate in themselves for an evaluation of a syllogism in that figure. They do not dispense with the need for an evaluation in the light of the eight rules already discussed. Rather each notes a requirement for a given figure which cannot be ignored without breaking one of the eight rules.

Special Rules for the First Figure
 For a direct conclusion —
  1. The minor premise must be affirmative.
  2. The major premise must be universal.
 For an indirect conclusion —
  3. If the minor premise is affirmative, the major must be universal.
  4. If the major premise is affirmative, the conclusion must be particular.
  5. If the conclusion is negative, the minor premise must be universal.

1. For a direct conclusion in the first figure the minor premise must be affirmative. If the minor premise were negative, the conclusion would be negative (Rule 6) and its predicate would be universal. Moreover, since the universal predicate of the conclusion is the major term, a universal major term would be required in the major premise (Rule 2). As predicate the major term could be universal only if the major premise were negative. But this would involve two negative premises and thus violate Rule 5.

2. For a direct conclusion in the first figure the major premise must be universal. A particular major premise would have a particular subject, in this case the middle term. The predicate of the minor premise, also the middle term, would have to be universal or else the middle would be twice particular (Rule 3). A universal predicate for the minor premise would require a negative minor premise. But, as we have just seen, the minor premise must be affirmative.

3. For an indirect conclusion, if the minor premise is affirmative, the major premise must be universal. If the minor premise is affirmative, its predicate, which is the middle term, is particular. If the middle term is particular in the minor premise, it must be universal in the major (Rule 3). As the subject of the major premise a universal middle term would make for a universal major premise.

4. For an indirect conclusion, if the major premise is affirmative, the conclusion must be particular. If the major premise is affirmative, the major term, which is its predicate, is particular. The major term must, then, be particular in the conclusion (Rule 2). As the subject of the conclusion a particular major term would determine a particular conclusion.

5. For an indirect conclusion, if the conclusion is negative the minor premise must be universal. If the conclusion is negative, its predicate, the minor term, is universal. If the minor term is universal in the conclusion, it must be universal in the minor premise (Rule 2). As subject, a universal minor term would make the minor premise universal.

THE RULES FOR THE CATEGORICAL SYLLOGISM

Special Rules for the Second Figure
For any conclusion —
 1. One premise (and, of course, the conclusion) must be negative.
For a direct conclusion —
 2. The major premise must be universal.
For an indirect conclusion —
 3. The minor premise must be universal.

1. For any conclusion in the second figure one premise (and, of course, the conclusion) must be negative. In the second figure the middle term is a predicate in both premises. Consequently, one premise must be negative; otherwise the middle term will not be universal at least once (Rule 3).

2. For a direct conclusion in the second figure the major premise must be universal. If the major premise were particular, its subject, the major term, would be particular. But the major term, as predicate of a negative conclusion, must be universal in the conclusion. The major term cannot be universal in the conclusion and only particular in the major premise (Rule 2).

3. For an indirect conclusion in the second figure the minor premise must be universal. If the minor premise were particular, its subject, the minor term, would be particular. But the minor term must be universal in the conclusion because it is the predicate of the conclusion, which must, as we saw above, be a negative conclusion. The minor term cannot be universal in the conclusion and only particular in the minor premise (Rule 2).

Special Rules for the Third Figure
For a direct conclusion —
 1. The minor premise must be affirmative.
For an indirect conclusion —
 2. The major premise must be affirmative.
For any conclusion —
 3. The conclusion must be particular.

1. For a direct conclusion in the third figure the minor premise must be affirmative. If the minor premise were negative, the conclusion would have to be negative (Rule 6). If the conclusion were negative, its predicate, the major term, would be universal. If the major term were universal in the conclusion, it would have to be universal in the major premise (Rule 2). As predicate of the major premise the major term could be universal only if the major premise were negative. But this would involve two negative premises and rule out validity (Rule 5).

2. **For an indirect conclusion in the third figure the major premise must be affirmative.** If the major premise were negative, the conclusion would have to be negative (Rule 6). If the conclusion were negative, its predicate, the minor term, would be universal. If the minor term were universal in the conclusion, it would have to be universal in the minor premise (Rule 2). As predicate of the minor premise the minor term could be universal only if the minor premise were negative. But this would involve two negative premises and rule out validity (Rule 5).

3. **For any conclusion in the third figure the conclusion must be particular.** If the minor premise leading to a direct conclusion must be affirmative, as we have just seen, its predicate, the minor term, must be particular. If the minor term is particular in the minor premise, it must be particular in the conclusion (Rule 2). As subject of the conclusion a particular minor term constitutes a particular conclusion. The same is true in reference to the major term in a syllogism leading to an indirect conclusion.

III. The Valid Moods of the Categorical Syllogism

A student who knows the six (or eight) rules for any categorical syllogism can confidently and capably evaluate any categorical syllogism. In this section we shall determine which of the possible moods of the categorical syllogism yield valid conclusions. The student should not memorize the list of these moods with the idea that he can thereby mechanically eliminate any syllogism whose mood is not valid. Rather the student should apply the rules he has learned to each possible mood to test its validity. By doing this he will draw up a list of the valid moods of the categorical syllogism.

First of all, what are the various possibilities? A mood is determined by the disposition of the premises in terms of quantity and quality. Consequently there are sixteen possible moods for each figure, since both major and minor premises can be either *A, E, I,* or *O*. Each of the four different major premises admits of four different minor premises so that the total number of combinations of premises is sixteen. Each mood, at least theoretically, can also have either an *A, E, I,* or *O* conclusion. Multiplying the sixteen possible combinations of premises by the four possible kinds of conclusions we find that there are sixty-four combinations possible. Since there are three figures, each with the possibility of both direct and indirect conclusions, the total number of combinations is 384 (4 [major premises] × 4 [minor premises] × 4 [conclusions] × 3 [figures] × 2 [types of conclusions in each figure]). As a matter of fact only a small fraction of this total number is valid. Of the 128 possible combinations in each figure there are only four distinctly different valid moods leading to a direct conclusion in the first figure and only five leading to an indirect conclusion, four of each in the second figure, and six of each in the third figure. Thus, out of 384 possible combinations of propositions there are only 29 valid moods of the categorical syllogism.

We begin with the 384 possible combinations of major premise, minor premise, and conclusion. Let us see how many survive an evaluation in the light of the general rules for the propositions in the categorical syllogism. First of all, we can eliminate any combination which violates Rule 5 by including two negative premises. This eliminates the combinations with EE, EO, OE, and OO premises. Since these four sets of premises admit theoretically of any one of four conclusions and can be found in each of three figures in reference to both direct and indirect conclusions, we have succeeded in eliminating 96 of our 384 possibilities simply by the application of Rule 5. Next let us look among the remaining combinations for those which violate Rule 7 by including two particular premises. This group includes all those combinations with II, IO, and OI premises (OO already eliminated by way of Rule 5). The total number in this group is 72. Now let us eliminate all those combinations which violate Rule 4 by concluding negatively from two affirmative premises. This group includes AA, AI, and IA premises when they are coupled with either an E or an O conclusion (II already eliminated by Rule 4). The total number of combinations in this group is 36. Some of the remaining combinations violate Rule 6 by concluding affirmatively even though one premise is negative. These include the AE, AO, EA, EI, IE, and OA combinations (the IO and OI combinations already eliminated by Rule 7) when coupled with an A or an I conclusion. The total number here is 72. Finally, we can apply Rule 8 to eliminate from what remains all those with universal conclusions and one particular premise. This includes the AIA, AOE, EIE, IAA, IEE, and OAE combinations (OEE, OEA, OAA, AIE, AOA, EIA, EOA, EOE, IAE, and IEA all having been eliminated through the use of other rules). The total in this group numbers 36. Thus, through the application of the general rules for categorical syllogism, we have ruled out as invalid all but 72 of the 384 possible combinations we started with.

These 72 remaining combinations include AAA, AAI, AEE, AEO, AII, AOO, EAE, EAO, EIO, IAI, IEO, and OAO considered in reference to both direct and indirect conclusions in each of the three figures. If we apply the special rules for the different figures of the categorical syllogism to these combinations, we find it necessary to eliminate all but 35. For example, Rule 4 rules AAA out for an indirect conclusion in the first figure; Rule 1 rules AAA out for any conclusion in the second figure; and Rule 3 rules AAA out for any conclusion in the third figure. The legitimate combinations for each of the figures can be listed in this fashion:

	Direct	Indirect
First Figure	AAA AAI AII EAE EAO EIO	AAI AEO AII EAE IEO
Second Figure	AEE AEO AOO EAE EAO EIO	AEE AEO EAE EAO IEO OAO
Third Figure	AII EAO EIO IAI OAO AAI	AEO AII AOO IAI IEO AAI

Note that six pairs have been boxed together. For each boxed pair the conclusion in one of the combinations is related to the other conclusion as a subalternate. The premises are able to generate the universal conclusion and *a fortiori* the particular, since the particular is contained in the universal. Consequently, the combination with the particular conclusion is not a distinct mood. It is only a variation of the mood which is proportioned to the universal conclusion. Out of the original 384 possibly valid combinations, only 35 survive an evaluation in the light of the rules — and these 35 represent but 29 distinctly different valid moods, 14 leading to direct conclusions in the three figures and 15 leading to indirect conclusions.

Most logicians do not count the indirect moods of the second and third figures, considering them as variant expressions of the direct moods of these figures. Thus, they give only 19 distinctly different valid moods, 14 leading to direct conclusions in the three figures and 5 leading to an indirect conclusion in the first figure. There is justification for this position, since each of the indirect conclusions in the second and third figures can be reached directly if the premises are transposed. This transposition of premises involves no difficulties at first glance, since, to begin with, the middle term is in the same position in both premises. However, it does demand that what was originally the major term now be considered as the minor, and vice versa. In some cases this does not pose any difficulty, for in some instances neither extreme is in itself more apt than the other to be used as major term. Consider the following syllogism, for example:

Every genus is said of many differing in species.
But no species is said of many differing in species.
Therefore, no genus is a species.

This appears to be a second figure syllogism concluding indirectly, with *genus* as major term and *species* as minor term. We can legitimately consider it as such, or we can transpose its premises, consider *species* as major term and *genus* as minor term, and count it as a second figure syllogism concluding directly. There is nothing to rule out this second interpretation, for certainly neither *genus* nor *species* has any logical priority over the other. One is as apt to be taken as major term as the other, for each is equally (i.e., on the same level) a subjective part of the same logical whole. We are as free to consider one the major term as the other. However, this does not mean that we can dispense with the indirect moods in the second figure simply because a syllogism such as this one need not be looked upon as having an indirect mood. First of all, let us take the syllogism in question to be the equivalent of this second figure syllogism concluding directly:

No species is said of many differing in species.
But every genus is said of many differing in species.
Therefore, no genus is a species.

The rules of syllogistic procedure allow this conclusion and, with these very same premises looked upon in exactly this same way, a second conclusion: *No species is a genus.* This second conclusion, in the light of these premises, is an indirect conclusion. Thus, no matter whether we take the premises as originally given (in which case the original conclusion is indirect) or transpose them (in which case the simple converse of the original conclusion is a legitimate indirect conclusion), we find ourselves faced with the actuality of an indirect conclusion in the second figure. The same thing can be shown to be true for the third figure.

The case for the independent consideration of indirect moods in the second and third figures is even stronger if we take an example with extremes which are not indifferent to consideration as to major and minor status. For example:

> Every body is corruptible.
> But some body is not a plant.
> Therefore, some corruptible thing is not a plant.

This appears to be a third figure syllogism concluding indirectly. It is possible to transpose its premises and thus reduce it to a third figure syllogism concluding directly. *Plant* would then be considered to be the major term and *corruptible* the minor term. However, it is clear that *corruptible*, since it is an attribute of *plant*, is apt to be said of or denied of *plant* rather than the other way round. The conclusion of this syllogism concluding directly would be inappropriately an indirect proposition. It is more accurate to take this syllogism as one which concludes indirectly than to consider it the equivalent of a syllogism concluding directly. And what is true of this example in the third figure is true of others, both in the second and third figures. For these reasons we have chosen to consider the valid indirect moods not only of the first figure but of the second and third as well.

We have treated the indirect moods in all three figures. However, it is only fair to point out that these are most significant in the first figure, and especially for two of the five legitimate indirect moods of the first figure. We have seen that the premises yielding valid indirect conclusions in the second and third figures can be looked upon as giving these same conclusions directly once the premises are transposed.[2] This maneuver is not effective in the first figure. Once the premises in the first figure are transposed and the predicate of the conclusion is considered to be the

---

[2] However, we have also seen that this way of looking upon indirect conclusions in the second and third figures involves some difficulties, and that it is more accurate to admit a real distinction between direct and indirect conclusions in all figures.

major term, then the major term occupies a place in the conclusion and in the premises which would indicate that it is both greater in extension and less in extension than the minor term. This is a logical absurdity. However, the (indirect) conclusions in *AAI*, *AII*, and *EAE* can be looked upon as being generated *directly* from their premises in the sense that each is the converse of a direct conclusion. This is not the case with *AEO* and *IEO*. The conclusions in these two indirect moods of the first figure can be considered as generated from these premises only indirectly. Neither can be seen to be generated directly from transposed premises (as in the case of the valid indirect moods of the second and third figures) nor as a converse of a direct conclusion (as in the case of the other valid indirect moods in the first figure). There is no way of viewing these except as distinctly independent moods concluding indirectly in the first figure. This is the reason why the indirect moods in the first figure have not been ignored (though, as we have seen, they have sometimes been mistaken for moods of a fourth figure), while the indirect moods of the second and third figures have usually been passed over as insignificant.

IV. The Reduction of the Syllogism to the First Figure

We have seen that the first figure of the categorical syllogism is the most perfect form of syllogistic procedure. The two basic principles of syllogistic discourse directly apply only to the first figure concluding directly. The first figure remains the only independent figure. The other figures are legitimate inasmuch as each can be reduced to the first. We can, of course, evaluate these figures directly by applying the eight rules which flow from the two basic principles without reducing them to a valid mood in the first figure. Still it is important to know how to reduce the moods of the second and third figures to a valid mood in the first. Only when this is done can we see clearly and immediately how the two basic principles of syllogistic reasoning are realized in any syllogism other than a syllogism in the first figure concluding directly.

There are two ways of doing this, *direct* and *indirect*. *Direct reduction* operates by converting and/or transposing the propositions of the original syllogism. *Indirect reduction* appeals to the principle of contradiction to show the impossibility of supposing that the mood in question is invalid. In direct reduction a syllogism on a par with the original is constructed in the first figure concluding directly. In indirect reduction a first figure syllogism concluding directly is constructed to show the absurdity of holding that the original syllogism is invalid.

Many reductions require relatively little effort. For example, any second figure syllogism with a direct conclusion whose major premise can be converted simply is directly reduced to a legitimate mood of the first figure by so converting it. The same thing is true of any third figure syllogism concluding directly as far as its minor premise is concerned. To illustrate:

No A is B.
But every C is B.
Therefore, no C is A. } → { No B is A.
But every C is B.
Therefore, no C is A.
(Second figure concluding directly with an E major)

Every B is A.
But some B is C.
Therefore, some C is A. } → { Every B is A.
But some C is B.
Therefore, some C is A.
(Third figure concluding directly with an I minor)

Another relatively simple reduction is possible in the case of several of the valid moods of the first figure concluding indirectly. Whenever a conclusion of this kind is simply convertible, the syllogism can be reduced to a mood of the first figure concluding directly merely by making the simple conversion of its conclusion. Thus:

No B is A.
But every C is B.
Therefore, no A is C. } → { No B is A.
But every C is B.
Therefore, no C is A.
(First figure concluding indirectly with an E conclusion)

Other direct reductions are more complicated. For example, an IAI syllogism concluding directly in the third figure can be directly reduced to a first figure concluding directly only through the simple conversion of both its major premise and its conclusion along with the transposition of its premises. Thus:

Some B is A.
But every B is C.
Therefore, some C is A. } → { Every B is C.
But some A is B.
Therefore, some A is C.
(Third figure concluding directly with IAI combination of propositions)

The AAI mood in the third figure concluding indirectly admits of two direct reductions. It can be reduced either by converting its major premise accidentally and transposing its premises or by converting its minor premise accidentally and converting its conclusion simply. To illustrate:

$$\left.\begin{array}{l}\text{Every } B \text{ is } A.\\ \text{But every } B \text{ is } C.\\ \text{Therefore, some } A \text{ is } C.\end{array}\right\} \rightarrow \left\{\begin{array}{l}\text{Every } B \text{ is } C.\\ \text{But some } A \text{ is } B.\\ \text{Therefore, some } A \text{ is } C.\end{array}\right.$$

(Third figure concluding indirectly with AAI propositions)

$$\left.\begin{array}{l}\text{Every } B \text{ is } A.\\ \text{But every } B \text{ is } C.\\ \text{Therefore, some } A \text{ is } C.\end{array}\right\} \rightarrow \left\{\begin{array}{l}\text{Every } B \text{ is } A.\\ \text{But some } C \text{ is } B.\\ \text{Therefore, some } C \text{ is } A.\end{array}\right.$$

All the valid moods which are not already in the first figure concluding directly can be indirectly reduced to it. Four of them can be reduced only through the method of indirect reduction. Indirect reduction is achieved in this way: first, contradict the conclusion of the original syllogism; then, use this as a premise in a first figure syllogism concluding directly along with the appropriate premise of the original syllogism to establish the denial of the other premise. The original syllogism is valid if the contradiction of its conclusion coupled with one of its original premises contradicts its other premise. A syllogism is valid precisely because its premises cannot be true if its conclusion is false. We can show that this is the status of any valid syllogism by assuming the opposite situation and then showing its absurdity. If we assume that the conclusion of any syllogism is false while its premises are true, then we assume the truth of the contradiction of that conclusion. If this contradiction is true, then it cannot be coupled with one of the premises we have assumed as true to establish the falsity of the other premise, also assumed as true. Whenever the contradiction of the conclusion and one of the premises do establish the falsity of the other premise, it is clear that we have an impossible situation. Thus the original conclusion could not have been false. It must have been true, and the original syllogism must have been valid.

We have seen how an *EAE* syllogism in the second figure concluding directly can be reduced through the simple conversion of its major premise. This same syllogism can be indirectly reduced to the first figure by contradicting its conclusion and using this as a premise along with its major premise to establish the contradiction of its minor premise:

$$\left.\begin{array}{l}\text{No } A \text{ is } B.\\ \text{But every } C \text{ is } B.\\ \text{Therefore, no } C \text{ is } A.\end{array}\right\} \rightarrow \left\{\begin{array}{l}\text{No } A \text{ is } B.\\ \text{But some } C \text{ is } A.\\ \text{Therefore, some } C \text{ is not } B.\end{array}\right.$$

An example of a valid syllogism in an imperfect mood which can be reduced to a perfect mood only by way of indirect reduction is the AOO syllogism in the second figure concluding directly. This reduction takes place as follows:

$$\left. \begin{array}{l} \text{Every } A \text{ is } B. \\ \text{But some } C \text{ is not } B. \\ \text{Therefore, some } C \text{ is not } A. \end{array} \right\} \rightarrow \left\{ \begin{array}{l} \text{Every } A \text{ is } B. \\ \text{But every } C \text{ is } A. \\ \text{Therefore, every } C \text{ is } B. \end{array} \right.$$

It is not necessary here to illustrate each reduction. However, it is a worthwhile exercise for the student to reduce directly each imperfect mood to a mood in the first figure concluding directly and, where this is impossible, to do so by an indirect reduction. The steps to be taken for each reduction are as follows:

First figure concluding indirectly	AAI — simply convert the conclusion. AEO — accidentally convert the major, simply convert the minor, and transpose the premises. AII — simply convert the conclusion. EAE — simply convert the conclusion. IEO — simply convert both premises and transpose the premises.
Second figure concluding directly	AEE — simply convert the minor, simply convert the conclusion, and transpose the premises. AOO — indirect reduction only. EAE — simply convert the major. EIO — simply convert the major.
Second figure concluding indirectly	AEE — simply convert the minor and transpose the premises. EAE — simply convert the major and simply convert the conclusion. IEO — simply convert the minor and transpose the premises. OAO — indirect reduction only.
Third figure concluding directly	AII — simply convert the minor. EAO — accidentally convert the minor. EIO — simply convert the minor. IAI — simply convert the major, simply convert the conclusion, and transpose the premises. OAO — indirect reduction only. AAI — accidentally convert the minor.
Third figure concluding indirectly	AEO — accidentally convert the major and transpose the premises. AII — simply convert the minor and simply convert the conclusion. AOO — indirect reduction only. IAI — simply convert the major and transpose the premises. IEO — simply convert the major and transpose the premises. AAI — accidentally convert the major and transpose the premises or accidentally convert the minor and simply convert the conclusion.

THE RULES FOR THE CATEGORICAL SYLLOGISM

Most logic textbooks suggest a timeworn set of mnemonic verses in Latin hexameter to help the student to recall the valid moods of the categorical syllogism and the method of reducing the various imperfect moods to perfect moods. Because Latin hexameters are, unfortunately, of little value as an aid to study for the modern student and because these traditional verses omit reference at least to some of the indirect moods, we shall spend little time on them. However, it may be at least of historical importance to note the names given the valid moods leading to direct conclusions and to note how the spelling of each name reveals the pattern of the mood and, in the case of the second and third figures, the method of reduction to the first figure. The names given the valid moods in the first figure are:

   Barbara   Darii
   Celarent   Ferio

Those in the second figure are called:

   Cesare   Festino
   Camestres   Baroco

In the third figure the valid moods are known as:

   Darapti   Felapton
   Disamis   Bocardo
   Datisi   Ferison

In each instance the vowels represent in order the major premise, minor premise, and conclusion. Thus, *Barbara* indicates an *AAA* mood, and *Bocardo* an *OAO* mood. The initial letter for the valid moods in the second and third figures indicates the mood in the first figure to which it must be reduced. Thus, Caesare is reduced to Celarent, Datisi to Darii, and Ferison to Ferio. An *s* (for *simplex*) indicates that the proposition for which the preceding vowel stands must be simply converted in reduction. A *p* (for *per accidens*) indicates that the preceding vowel stands for a proposition which must be accidentally converted. An *m* (for *mutatio*) in the name indicates that the premises must be transposed. Finally a *c* (for *contradictio*) indicates that the only possible reduction is an indirect reduction by way of contradiction. It should be of interest to the student to check the accuracy of these names and, in the event they can be helpful, to use them as a mnemonic aid in recalling the valid moods of the categorical syllogism.

Exercise XVI

1. Differentiate between the basic principles for the categorical syllogism, the general rules for the categorical syllogism, and the special rules for the different figures of the categorical syllogism.
2. State each of the eight general rules for the categorical syllogism, and defend each one in the light of the basic principles.
3. State each of the special rules for each figure, and defend each one in the light of the general rules.
4. Explain why it is that there are 384 possible combinations of propositions for the categorical syllogism. Show how all but 35 of these are rejected as invalid. Explain why it is that only 29 of these 35 are considered to be distinct moods.
5. Some logicians speak of 19 valid moods in four figures. Others speak of 19 valid moods in three figures. We have spoken of 29 valid moods in three figures. What can be said for and/or against each position?
6. What is the significance of the reduction of second and third figure syllogisms and first figure syllogisms concluding indirectly to first figure syllogisms concluding directly?
7. How does indirect reduction differ from direct reduction?
8. Construct an example for each valid mood in the second and third figure and in the first figure concluding indirectly and reduce each one directly and/or indirectly to a mood of the first figure concluding directly.
9. Try to construct mnemonic names for the valid indirect moods in the three figures, using those for the direct moods as models.
10. Classify each of the following syllogisms by indicating its figure and by pointing out whether its conclusion is direct or indirect. Then evaluate it as valid or invalid, and, if it is invalid, explain what is wrong with it. Reduce each valid syllogism to a valid mood of the first figure concluding directly:
    10.1. No crime is perfect.
          But every crime is an immorality.
          Therefore, no immorality is perfect.
    10.2. Every virtue is worth having.
          But some habits are not virtues.
          Therefore, some habits are not worth having.
    10.3. No creature is necessary.
          But no creature is God.
          Therefore, God is necessary.
    10.4. Every horse is a substance.
          But every horse is a brute animal.
          Therefore, some substance is a brute animal.
    10.5. No man is a hippopotamus.
          But some man is virtuous.
          Therefore, some hippopotamus is not virtuous.

THE RULES FOR THE CATEGORICAL SYLLOGISM 219

10.6. The philosophy teachers teach all the logic classes in this school.
But Professor Allen is a philosophy teacher.
Therefore, Professor Allen teaches all the logic classes in this school.

10.7. No polygons are circles.
But some figures are not circles.
Therefore, some figures are not polygons.

10.8. Every definition is a composite expression.
But every definition is convertible with its subject.
Therefore, every composite expression is convertible with its subject.

10.9. No news is good news.
But the end of a war is news.
Therefore, the end of a war is not good news.

10.10. Every .300 hitter is a help to his baseball team.
But Jamie is not a .300 hitter.
Therefore, Jamie is not a help to his baseball team.

10.11. Every polygon is rectilinear.
But every polygon is a plane figure.
Therefore, some plane figure is rectilinear.

10.12. Every virtue is a quality.
But every ability is a quality.
Therefore, every ability is a virtue.

10.13. Some liquid is inflammable.
But some liquid is kerosene.
Therefore, some kerosene is inflammable.

10.14. Every moral agent is free.
But no plant is free.
Therefore, no moral agent is a plant.

11. Supply whatever direct and indirect conclusions are legitimate for the following premises. Classify each completed syllogism according to figure and mood:

11.1. Every A is B.
But no A is C.
Therefore,

11.2. Every A is B.
But no C is B.
Therefore,

11.3. Some A is B.
But every A is C.
Therefore,

11.4. Every A is B.
But some C is B.
Therefore,

11.5. No A is B.
But every C is B.
Therefore,

11.6. Some A is B.
But no C is A.
Therefore,

11.7. Some A is not non-B.
But every C is A.
Therefore,

11.8. Every A is non-B.
But no C is B.
Therefore,

11.9. Some A is not B.
But every A is C.
Therefore,

11.10. Some A is not B.
But no A is C.
Therefore,

12. Restate each of the following syllogisms in strictly logical form; then classify each as to figure and mood, and evaluate each in the light of the rules:

12.1. Since you must believe what you can plainly see, you must believe me, because you can plainly see me.

12.2. Labor unions should be abolished since all unjust organizations should be abolished, and some labor unions are unjust organizations.

12.3. John is a humble man, because he knows his own weaknesses, and every humble man knows his own weaknesses.

12.4. No circle is a polygon, because no circle is quadrilateral, and some polygon is quadrilateral.

12.5. No prudent politician will accept expensive gifts, and no one who accepts expensive gifts can escape suspicion and criticism. Hence, some prudent politicians escape suspicion and criticism.

12.6. Some men are not patriotic, because some men are not loyal Americans, and loyal Americans are patriotic.

12.7. Capital punishment is forbidden by the Bible, because to inflict capital punishment is to kill, and one of the commandments of God recorded in the Bible forbids killing.

12.8. Some scientists are not rigorously logical, for all scientists are men, and not all men are rigorously logical.

12.9. Because it is a book, *Lady Chatterley's Lover* is one of the greatest influences for spiritual good in the history of man; for books are most certainly one of the greatest influences for spiritual good in the history of man.

12.10. Some drugs are habit-forming, and not all drugs are of medicinal value. Hence, not all habit-forming substances are of medicinal value.

13. Restate in strictly logical form the syllogisms in the following passage from St. Thomas Aquinas, classifying each as to figure and mood, and evaluating each in the light of the rules for syllogistic procedure:

"From this, moreover, it is also clear that riches are not the highest good for man.

"Indeed, riches are only desired for the sake of something else; they provide no good of themselves but only when we use them, either for the maintenance of the body or some such use. Now, that which is the highest good is desired for its own sake and not for the sake of something else. Therefore, riches are not the highest good for man.

"Again, man's highest good cannot lie in the possession or keeping of things that chiefly benefit man through being spent. Now, riches are chiefly valuable because they can be expended, for this is their use. So, the possession of riches cannot be the highest good for man.

"Besides, an act of virtue is praiseworthy in so far as it comes closer to felicity. Now, acts of liberality and magnificence, which have to do with money, are more praiseworthy in a situation in which money is spent than in one in which it is saved. So, it is from this fact that the names of these virtues are derived. Therefore, the felicity of man does not consist in the possession of riches.

"Moreover, that object in whose attainment man's highest good lies must be better than man. But man is better than riches, for they are but things subordinated to man's use. Therefore, the highest good for man does not lie in riches.

"Furthermore, man's highest good is not subject to fortune, for things subject to fortune come about independently of rational effort. But it must be through reason that man will achieve his proper end. Of course, fortune occupies an important place in the attainment of riches. Therefore, human felicity is not founded on riches.

"Again, this becomes evident in the fact that riches are lost in an involuntary manner, and also that they may accrue to evil men who must fail to achieve the highest good and also that riches are unstable — and for other reasons of this kind which may be gathered from the preceding arguments."

(St. Thomas Aquinas, *Contra Gentiles*, Bk. III, Chap. 30. Translation by V. Bourke, *On the Truth of the Catholic Faith*, Bk. III, Pt. I [Garden City, N. Y.: Image Books, 1956], pp. 116–117.) Reprinted with permission of Doubleday and Co.

14. Look for examples of syllogistic argumentation in the textbooks and reference works used in this course and in other courses. Restate each syllogism in strictly logical form, and classify and evaluate each of them.

CHAPTER XVII

The Hypothetical Syllogism

I. The Nature and Types of the Hypothetical Syllogism

In our investigation into the division of the proposition we spent a good deal of time and effort on the hypothetical proposition. One reason for this is the fact that hypothetical propositions can be employed as major premises in a distinctive type of deductive argumentation called the hypothetical syllogism. In this chapter we will investigate the nature and types of hypothetical syllogism. Since our consideration here will presume all that we saw previously in reference to the hypothetical proposition, the student is advised to review carefully the chapter on the hypothetical proposition.

A hypothetical proposition is a compound proposition which involves a commitment to a nexus or link between its component propositions without asserting the truth status of any one of them.

The component propositions in a hypothetical proposition are so related that the truth status of any one of them has a bearing on the truth status of the others. For this reason, once the intellect sees the truth or falsity of any one of the components of a hypothetical proposition, it is moved to affirm or deny the other element or elements. For example, the truth of *John is a Communist*, in the hypothetical proposition *If John is a Communist, his children are in spiritual danger*, forces assent to the truth of the proposition *John's children are in spiritual danger*. Here we have a clear-cut case of hypothetical reasoning which can be expressed in the following argument:

> If John is a Communist, his children are in spiritual danger.
> But John is a Communist.
> Therefore, John's children are in spiritual danger.

The simple conditional premise in this syllogism asserts that its elements are so related that if the first is true the second must also be true. The categorical premise asserts the truth of the first element of the conditional.

The truth of the second element of the conditional follows as a categorical conclusion. This is an instance of a hypothetical syllogism. The hypothetical premise functions as major premise and the categorical premise functions as minor precisely because the categorical premise is taken under the hypothetical and contracts it, in a sense, to the conclusion. There are, of course, no major, minor, and middle terms. Hypothetical argument works in virtue of the affirmation or denial of sequentially connected propositions and not, as is the case with the categorical syllogism, in virtue of the triple identity of terms.

Every hypothetical syllogism can be resolved into three propositions: the major premise, the minor premise, and the conclusion. The major premise is a hypothetical proposition. The minor premise is a proposition (either categorical or compound) which affirms or denies one or more of the elements of the major. The conclusion is a proposition (either categorical or compound) which affirms or denies the other elements of the major. The simplest and most frequently employed form of hypothetical argumentation has a two-membered hypothetical major, a categorical minor affirming or denying one of the two elements of the major, and a categorical conclusion affirming or denying the other element. The example suggested above is an instance of this relatively simple form of hypothetical syllogism. Throughout this chapter we shall limit ourselves for the most part to a consideration of this elementary type of hypothetical syllogism, reserving an investigation into syllogisms with multimembered majors on the one hand and majors with compound components on the other to a special section of the chapter.

We have seen that there are five types of hypothetical proposition, namely, the simple conditional, the reciprocal conditional, the inclusive alternative, the exclusive alternative, and the disjunctive. Each of these can be used as a major premise in a hypothetical syllogism. Each specifies a different type of syllogism with a different set of rules. We shall take up each type of hypothetical syllogism separately, carefully discerning the special set of rules proper to each.

II. The Figures and Moods of the Hypothetical Syllogism

There are theoretically two different figures for each type of hypothetical syllogism, although some hypothetical syllogisms admit of a valid conclusion in only one of these two figures. The figure of the hypothetical syllogism is determined by the minor premise. The minor premise either affirms or denies an element of the major. When the minor premise affirms an element of the major, the syllogism is said to be in the first figure. When the minor denies an element of the major, the syllogism is in the second figure. Thus, it is relatively easy to determine the figure of any hypothetical syllogism.

However, a word of caution is necessary. Students are easily misled into thinking that the figure of a hypothetical syllogism depends absolutely upon the quality of the minor premise. Some think that an affirmative minor as such constitutes the first figure and that a negative minor as such constitutes the second figure. This is not the case. The minor considered *absolutely* does not determine the figure of the syllogism. It is the minor in its determinate *reference* to some element of the major which determines the figure of the hypothetical syllogism. An affirmative minor determines the first figure only if it represents an affirmation of some one of the component parts of the major. An affirmative minor bearing upon a negative element of the major constitutes a denial of that element of the major and determines a second figure syllogism. In similar fashion a negative minor gives the second figure only if it bears upon an affirmative element of the major premise. A negative minor gives the first figure whenever it bears upon a negative element of the major. The affirmation of an affirmative proposition is affirmative. Its denial is negative. The denial of a negative proposition is affirmative, and its affirmation is negative.

A minor premise in a hypothetical syllogism sufficiently affirms an element of the major only if it affirms that element in its entirety. Thus an A element cannot be considered affirmed on the basis of a minor premise asserting the truth of the corresponding I or A^1. As a matter of fact, the truth of the I or the A^1 is compatible with the truth of the corresponding O, which is the contradiction of the A. The minor premise in a proposed hypothetical syllogism can legitimately constitute a first figure syllogism only if it represents a total affirmation of an element of the major, precisely because anything less than this is compatible with the contradiction of that element. There is, of course, no reason why one could not conclude to an I proposition in place of an A so long as the premises warranted the A, or to an O in place of an E. Here, as in the case of the categorical syllogism, the inferior conclusion is contained in the superior and is allowed wherever the superior is legitimate.

A minor premise in a second figure hypothetical syllogism functions adequately so long as it represents at least a simple denial of an element of the major premise. The minor premise in a second figure hypothetical syllogism can be either the contradiction or the contrary of the element it denies. However, when the premises in a hypothetical syllogism demand a conclusion which denies an element of the major, that conclusion can only be the contradiction of the element it denies or the inferior of that contradiction. The contrary of a proposition is more than a simple denial because it denies not only that proposition but its inferior as well, and the premises of a hypothetical proposition are able to generate nothing more than the affirmation or simple denial of some element of the major premise. The contrary can be used as a minor premise in so far as it involves *at least* a simple denial. The contrary cannot be used as a conclusion because it involves more than a simple denial.

The moods of the hypothetical syllogism depend upon the quality of the component propositions in the hypothetical major premise. Two-membered majors allow for four moods, while multimembered majors allow for proportionately more. In a two-membered major both elements can be affirmative, both negative, the first affirmative and the second negative, or the first negative and the second affirmative. These four combinations constitute the four moods possible for any hypothetical syllogism with a two-membered major premise. The moods of the hypothetical syllogism are not as significant as those of the categorical syllogism, and little reference will be made to them from now on.

As an illustration of differences in figure and mood note the following four hypothetical syllogisms. Each is valid. The first and third are simple conditional syllogisms. The second is an exclusive alternative. The fourth is a reciprocal conditional. The first two are in the first figure. The last two are in the second figure. The first three are in different moods.

1. If the weather is bad, the game is off.
   But the weather is bad.
   Therefore, the game is off.
2. Either the weather is bad or the game is not off, but not both.
   But the game is not off.
   Therefore, the weather is not bad.
3. If the weather is not bad, the game is on.
   But the game is not on.
   Therefore, the weather is bad.
4. If and only if the weather is not bad, the game is on.
   But the weather is bad.
   Therefore, the game is not on.

III. The Simple Conditional Syllogism

The fundamental type of hypothetical proposition is the conditional. Similarly, the fundamental type of hypothetical syllogism is the conditional. The simple conditional syllogism is a hypothetical syllogism with a simple conditional major premise and a minor premise which affirms or denies one of the elements of the major. The conclusion of the simple conditional syllogism affirms or denies the other element of the major according to the demands of the special rules applying to the simple conditional syllogism:

> An affirmation of the antecedent requires an affirmation of the consequent, and a denial of the consequent requires a denial of the antecedent.
>
> No conclusion follows from a denial of the antecedent or from an affirmation of the consequent.

The reason for these rules is found in the nature of the simple conditional proposition. The simple conditional proposition is defined in terms of a sequential nexus between its antecedent and its consequent. It asserts that its antecedent cannot be true without its consequent being true. It does not assert any reciprocal sequential nexus between its consequent and its antecedent. Thus the simple conditional proposition demands that its consequent be true whenever its antecedent is true. But it allows for the truth of its consequent even if its antecedent is false. This means that nothing follows either from the affirmation of its consequent or from the denial of its antecedent. The consequent can be true with a true or false antecedent, and the antecedent can be false with a true or false consequent. The falsity of the antecedent, however, follows from the falsity of the consequent. If the consequent could be false and the antecedent true, then the simple conditional could not assert, as it does, that the truth of the antecedent necessitates the truth of the consequent.

The first two of the following four syllogisms are invalid because they break the rules for the simple conditional syllogism. The last two are the only valid syllogisms which could be constructed with the major premise which has been used as common to all four.

1. If Freud's psychological theory is completely correct, then man is not a morally responsible agent.
   But man is not a morally responsible agent.
   Therefore, Freud's psychological theory is completely correct.
   (*Invalid:* Nothing follows from the affirmation of the consequent in a simple conditional syllogism.)
2. If Freud's psychological theory is completely correct, then man is not a morally responsible agent.

But Freud's psychological theory is not completely correct.
Therefore, a man is a morally responsible agent.
(*Invalid:* Nothing follows from the denial of the antecedent in a simple conditional syllogism.)
3. If Freud's psychological theory is completely correct, then man is not a morally responsible agent.
But Freud's psychological theory is completely correct.
Therefore, man is not a morally responsible agent.
(*Valid:* The affirmation of the consequent *does* follow from the affirmation of the antecedent.)
4. If Freud's psychological theory is completely correct, then man is not a morally responsible agent.
But man is a morally responsible agent.
Therefore, Freud's psychological theory is not completely correct.
(*Valid:* The denial of the antecedent *does* follow from the denial of the consequent.)

IV. The Case of the "Disguised" Categorical Major

In the chapter on the compound proposition we spoke about propositions that could be understood as categorical although expressed grammatically in hypothetical form. They should, we said, be taken as categorical rather than hypothetical, because they function better as major premises in syllogisms which follow the rules of categorical rather than hypothetical argumentation. Take the proposition *If anyone is disposed to act unjustly, he will make an uncertain friend*. This can be coupled with an A, A¹, or I minor affirming its "antecedent" and yielding respectively an A, A¹, or I conclusion affirming its "consequent." Thus each of the following is a valid argumentation:

1. If anyone is disposed to act unjustly, he will make an uncertain friend.
   But every man is disposed to act unjustly.
   Therefore, every man will make an uncertain friend.
2. If anyone is disposed to act unjustly, he will make an uncertain friend.
   But some men are disposed to act unjustly.
   Therefore, some men will make uncertain friends.
3. If anyone is disposed to act unjustly, he will make an uncertain friend.
   But John is disposed to act unjustly.
   Therefore, John will make an uncertain friend.

However, hypothetical argument requires the minor in a first figure syllogism to affirm an element of the major *in its entirety*. Thus syllogisms which validly admit A, A^1, and I propositions as legitimate minor premises (when the antecedent of the major is not clearly an I proposition) for the same major premise cannot *without difficulty* be taken as hypothetical argumentations. The three syllogisms above can be better understood to work if their common major premise is understood as *Everyone who is disposed to act unjustly makes an uncertain friend* rather than the corresponding conditional. With this as their major premise the three syllogisms can be readily seen to be respectively legitimate AAA, AII, and AA^1A^1 categorical syllogisms in the first figure. The rules for the categorical syllogism clearly allow for the passage from the universal major to the particular or singular conclusion by way of a particular or singular minor. The rules for the hypothetical syllogism, however, demand *total* affirmation for any minor which is taken as the affirmation of some element of the major. The point is not that "If anyone is disposed to act unjustly, he will make an uncertain friend" *cannot* be taken as a hypothetical proposition. The point is that it *can* be understood as a categorical proposition and that it is ordinarily preferable to take it in this way.

In our chapter on the compound proposition we noted that these "disguised" categoricals can take the form of alternatives and disjunctives as well as conditionals. There are apparent alternative and disjunctive syllogisms as well as apparent conditional syllogisms which are better understood as expressions of categorical syllogisms. Each of the following syllogisms might be more clearly expressed as categorical syllogisms:

1. *First expression:*
    Either a man is virtuous or he is not truly happy.
    But John is truly happy.
    Therefore, John is virtuous.
   Second expression:
    Every man who is truly happy is virtuous.
    But John is a man who is truly happy.
    Therefore, John is virtuous.
2. *First expression:*
    One cannot be both a miser and a spendthrift.
    But some men are misers.
    Therefore, some men are not spendthrifts.
   Second expression:

No one who is a miser is a spendthrift.
But some men are misers.
Therefore, some men are not spendthrifts.

Before proceeding to the second type of hypothetical syllogism it should be of value to show how simple conditional syllogisms can be represented symbolically. Here as elsewhere, when the concern is one of formal logic, symbols can be of great assistance. The student is referred to the chapter on hypothetical propositions, where the symbolic representation of the hypotheticals is explained and put into practice. It is important to recall that the literal symbols in the hypothetical proposition stand for categorical propositions and not for terms as they do in the categorical proposition itself. The following simple conditional syllogism can be accurately and conveniently symbolized in either of the ways illustrated:

If children are characteristically free of discipline in the home, juvenile delinquency can be expected to be the rule.
But children are characteristically free of discipline in the home.
Therefore, juvenile delinquency can be expected to be the rule.

$a \rightarrow b$ or $[(a \rightarrow b) \cdot a] \rightarrow b$
But a
Therefore, b

V. The Reciprocal Conditional Syllogism

The reciprocal conditional syllogism is a hypothetical syllogism with a reciprocal conditional major premise and a minor premise which affirms or denies one of the elements of the major. The conclusion of the reciprocal conditional syllogism affirms or denies the other element of the major according to the demands of the special rule applying to the reciprocal conditional syllogism:

An affirmation of either element requires the affirmation of the other element, and a denial of either element requires the denial of the other.

The reciprocal conditional proposition by definition involves a reciprocal conditional connection between its elements. Its antecedent cannot be true without its consequent being true, and its consequent cannot be true without its antecedent being true. Thus, if either one is true, the other must be true, and if either one is false, the other must be false.

Neither element can be true while the other is false because the truth of either implies the truth of the other. Neither can be false while the other is true; otherwise the truth of either would not be, as the reciprocal conditional asserts, incompatible with the falsity of the other. The difference between the simple conditional proposition and the reciprocal conditional proposition is that the former can have a valid conclusion only if its antecedent is affirmed or its consequent denied whereas the latter gives valid conclusions if either of its elements — antecedent or consequent — is affirmed or denied.

The following four syllogisms illustrate valid reciprocal conditional syllogisms. Since they share a common major premise they illustrate the four different conclusions virtually contained in any reciprocal conditional proposition.

1. If, and only if, the invasion is successful, the reputation of the commanding general is secured.
   But the invasion is successful.
   Therefore, the reputation of the commanding general is secured.
   (*Valid:* The affirmation of the consequent does follow from the affirmation of the antecedent in a reciprocal conditional syllogism.)

2. If, and only if, the invasion is successful, the reputation of the commanding general is secured.
   But the reputation of the commanding general is secured.
   Therefore, the invasion is successful.
   (*Valid:* The affirmation of the antecedent does follow from the affirmation of the consequent.)

3. If, and only if, the invasion is successful, the reputation of the commanding general is secured.
   But the invasion is not successful.
   Therefore, the reputation of the commanding general is not secured.
   (*Valid:* The denial of the consequent does follow from the denial of the antecedent.)

4. If, and only if, the invasion is successful, the reputation of the commanding general is secured.
   But the reputation of the commanding general is not secured.
   Therefore, the invasion is not successful.
   (*Valid:* The denial of the antecedent does follow from the denial of the consequent.)

Symbolically these four syllogisms would appear as:

1. $a \leftrightarrow b$ or $[(a \leftrightarrow b) \cdot a] \rightarrow b$
   But a
   Therefore, b

2. $a \leftrightarrow b$ or $[(a \leftrightarrow b) \cdot b] \rightarrow a$
   But b
   Therefore, a

3. $a \leftrightarrow b$ or $[(a \leftrightarrow b) \cdot -a] \rightarrow -b$
   But $-a$
   Therefore, $-b$

4. $a \leftrightarrow b$ or $[(a \leftrightarrow b) \cdot -b] \rightarrow -a$
   But $-b$
   Therefore, $-a$

VI. The Inclusive Alternative Syllogism

The inclusive alternative syllogism is a hypothetical syllogism with an inclusive alternative major premise and a minor premise which denies one of the elements of the major. Its conclusion affirms the other element of the major according to the special rules applying to the inclusive alternative syllogism:

The affirmation of either element yields no conclusion.

The denial of either element necessitates the affirmation of the other.

The nexus asserted between the elements of an inclusive alternative proposition is such that if one is false the other must be true. It asserts no more than this, allowing for the possibility that the elements may both be true. Thus the affirmation of either element in the minor leaves the conclusion in doubt, since the other element may or may not be true. But the denial of one of the elements of the major yields a conclusion which affirms the other, for the elements cannot both be false. Hence, the inclusive alternative syllogism allows for a conclusion in the second figure, but it allows for no conclusion in the first figure.

Each of the following four syllogisms is an example of an inclusive alternative syllogism. The first two are in the first figure and hence are invalid. The last two are in the second figure, and they are valid.

1. Either Aristotle and Plato were good philosophers or modern Scholasticism is academically unsound.

But Aristotle and Plato were good philosophers.
Therefore, modern Scholasticism is not academically unsound.
(*Invalid:* No conclusion follows in the first figure of an inclusive alternative syllogism.)

2. Either Aristotle and Plato were good philosophers or modern Scholasticism is academically unsound.
But modern Scholasticism is academically unsound.
Therefore, Aristotle and Plato were not good philosophers.
(*Invalid:* No conclusion follows in the first figure.)

3. Either Aristotle and Plato were good philosophers or modern Scholasticism is academically unsound.
But Aristotle and Plato were not good philosophers.
Therefore, modern Scholasticism is academically unsound.
(*Valid:* The denial of one element necessitates the affirmation of the other.)

4. Either Aristotle and Plato were good philosophers or modern Scholasticism is academically unsound.
But modern Scholasticism is not academically unsound.
Therefore, Aristotle and Plato were good philosophers.
(*Valid:* The denial of either element necessitates the affirmation of the other.)

These four syllogisms can be symbolically represented as follows:

1. $a \lor b$ or $[(a \lor b) \cdot a] \to -b$
But a
Therefore, $-b$

2. $a \lor b$ or $[(a \lor b) \cdot b] \to -a$
But b
Therefore, $-a$

3. $a \lor b$ or $[(a \lor b) \cdot -a] \to b$
But $-a$
Therefore, b

4. $a \lor b$ or $[(a \lor b) \cdot -b] \to a$
But $-b$
Therefore, a

VII. The Exclusive Alternative Syllogism

The exclusive alternative syllogism is a hypothetical syllogism with an exclusive alternative major premise and a minor premise which affirms or

THE HYPOTHETICAL SYLLOGISM 233

denies one of the elements of the major. The conclusion of the exclusive alternative syllogism denies or affirms the other element of the major according to the special rules for the exclusive alternative syllogism.

The affirmation of either element necessitates the denial of the other.

The denial of either element necessitates the affirmation of the other.

In an exclusive alternative proposition the connection between its elements is such that one must be true and any other false. Thus in a two-membered exclusive alternative the falsity of one element demands the truth of the other, and the truth of one demands the falsity of the other. If the minor affirms an element of the major, the conclusion must deny the other. If the minor denies an element of the major, the conclusion must affirm the other.

Each of the following four syllogisms is an example of a valid exclusive alternative syllogism. The first two are in the first figure, and the last two are in the second figure. Although the inclusive alternative syllogism works only in the second figure, the exclusive alternative allows for a conclusion in both figures.

1. Either logic is an indispensable tool for philosophy or philosophy is not rigorously scientific, but not both.
   But logic is an indispensable tool for philosophy.
   Therefore, philosophy is rigorously scientific.
   (*Valid:* The affirmation of one element necessitates the denial of the other.)

2. Either logic is an indispensable tool for philosophy or philosophy is not rigorously scientific, but not both.
   But philosophy is not rigorously scientific.
   Therefore, logic is not an indispensable tool for philosophy.
   (*Valid:* The affirmation of one element necessitates the denial of the other.)

3. Either logic is an indispensable tool for philosophy, or philosophy is not rigorously scientific, but not both.
   But logic is not an indispensable tool for philosophy.
   Therefore, philosophy is not rigorously scientific.
   (*Valid:* The denial of one element necessitates the affirmation of the other.)

4. Either logic is an indispensable tool for philosophy, or philosophy is not rigorously scientific, but not both.
But philosophy is rigorously scientific.
Therefore, logic is an indispensable tool for philosophy.
(*Valid:* The denial of one element necessitates the affirmation of the other.)

These four syllogisms can be presented symbolically in this fashion:

1. $a \wedge -b$ or $[(a \wedge -b) \cdot a] \to b$
But a
Therefore, b

2. $a \wedge -b$ or $[(a \wedge -b) \cdot -b] \to -a$
But $-b$
Therefore, $-a$

3. $a \wedge -b$ or $[(a \wedge -b) \cdot -a] \to -b$
But $-a$
Therefore, $-b$

4. $a \wedge -b$ or $[(a \wedge -b) \cdot b] \to a$
But b
Therefore, a

VIII. The Disjunctive Syllogism

The disjunctive syllogism is a hypothetical syllogism with a disjunctive major premise and a minor which affirms one of the elements of the major. The conclusion of a disjunctive syllogism denies the other element of the major according to the special rules for the disjunctive syllogism:

The affirmation of either element requires the denial of the other.

The denial of either element yields no conclusion.

The nexus between the component propositions of a disjunction is such that one at least must be false. This is the extent of the disjunctive commitment. It allows for the possibility that all of its components might be false. If any element is affirmed the other must be denied, for one at least must be false. Nothing can be concluded from the falsity of any element, since all can be false, but need not be.

Whereas the inclusive alternative works only in the second figure, the disjunctive syllogism works only in the first figure. The exclusive alternative, we saw, works in both the first and second figures. It is interesting to note that a two-membered exclusive alternative proposition can be

THE HYPOTHETICAL SYLLOGISM 235

resolved into a copulative proposition composed of corresponding inclusive alternative and disjunctive propositions. Thus, $a \wedge b \equiv (a \vee b) \cdot n - (a \cdot b)$. Along the same line, recall that the simple conditional syllogism works *sometimes* in the first figure and *sometimes* in the second figure, whereas the reciprocal conditional *always* works in both figures. It can do so because it can be resolved into a copulative proposition composed of simple conditionals with inverted antecedents and consequents. Thus: $a \leftrightarrow b \equiv (a \rightarrow b) \cdot (b \rightarrow a)$.

As an illustration of the disjunctive syllogism consider these four examples. The first two are in the first figure, and they are valid. The last two are in the second figure, and they are, accordingly, invalid.

1. It cannot be true both that Communism is a true philosophy and that Christianity is a true religion.
   But Communism is a true philosophy.
   Therefore, Christianity is not a true religion.
   (*Valid:* The affirmation of either element necessitates the denial of the other.)

2. It cannot be true both that Communism is a true philosophy and that Christianity is a true religion.
   But Christianity is a true religion.
   Therefore, Communism is not a true philosophy.
   (*Valid:* The affirmation of either element necessitates the denial of the other.)

3. It cannot be true both that Communism is a true philosophy and that Christianity is a true religion.
   But Communism is not a true philosophy.
   Therefore, Christianity is a true religion.
   (*Invalid:* Nothing follows from the denial of either element.)

4. It cannot be true both that Communism is a true philosophy and that Christianity is a true religion.
   But Christianity is not a true religion.
   Therefore, Communism is a true philosophy.
   (*Invalid:* Nothing follows from the denial of either element.)

Symbolically these syllogisms can be rendered as follows:

1. $n - (a \cdot b)$ or $[n - (a \cdot b) \cdot a] \rightarrow -b$
   But a
   Therefore, $-b$

2. $n - (a \cdot b)$ or $[n - (a \cdot b) \cdot b] \rightarrow -a$
   But b
   Therefore, $-a$

3. $n - (a \cdot b)$ or $[n - (a \cdot b) \cdot -a] \to b$
   But $-a$
   Therefore, b

4. $n - (a \cdot b)$ or $[n - (a \cdot b) \cdot -b] \to a$
   But $-b$
   Therefore, a

IX. The Reduction of Alternative and Disjunctive Syllogisms to the Conditional Syllogism

The conditional proposition, as we know, is the fundamental type of hypothetical proposition; to it alternative and disjunctive propositions can be reduced. Consequently, the conditional syllogism is the fundamental type of hypothetical syllogism and alternative and disjunctive syllogisms can be reduced to equivalent conditional syllogisms.

The inclusive alternative proposition can be reduced to an equivalent simple conditional. Thus, $a \lor b \equiv -a \to b$. Two valid syllogisms can be constructed with $a \lor b$ as the major premise. These are:

1. $a \lor b$ 2. $a \lor b$
   But $-a$ But $-b$
   Therefore, b Therefore, a

If the conditional equivalent of the major premise is combined with the same minor two equivalent simple conditional syllogisms can be formed. These are:

1. $-a \to b$ 2. $-a \to b$
   But $-a$ But $-b$
   Therefore, b Therefore, a

The disjunctive proposition is also reduced to a corresponding simple conditional. We saw $n - (a \cdot b) \equiv a \to -b$. The two valid disjunctive syllogisms which can be constructed with $n - (a \cdot b)$ as the major premise are:

1. $n - (a \cdot b)$ 2. $n - (a \cdot b)$
   But a But b
   Therefore, $-b$ Therefore, $-a$

The simple conditional syllogisms to which they can be reduced are:

1. $a \to -b$ 2. $a \to -b$
   But a But b
   Therefore, $-b$ Therefore, $-a$

THE HYPOTHETICAL SYLLOGISM

The exclusive alternative proposition is reduced to a reciprocal conditional. Thus, $a \wedge b \equiv a \leftrightarrow -b$; $a \wedge b$ admits of four valid conclusions in syllogisms which can be reduced to valid reciprocal conditional syllogisms with $a \leftrightarrow -b$ as the major premise. The four exclusive alternatives which can be constructed from $a \wedge b$ and the four reciprocal conditionals to which they can be reduced are as follows:

1. $a \wedge b$	2. $a \wedge b$	3. $a \wedge b$	4. $a \wedge b$
But a	But b	But $-a$	But $-b$
Therefore, $-b$	Therefore, $-a$	Therefore, b	Therefore, a
1. $a \leftrightarrow -b$	2. $a \leftrightarrow -b$	3. $a \leftrightarrow -b$	4. $a \leftrightarrow -b$
But a	But b	But $-a$	But $-b$
Therefore, $-b$	Therefore, $-a$	Therefore, b	Therefore, a

X. Hypothetical Syllogisms With a Multimembered Major

Thus far we have considered only hypothetical syllogisms with a two-membered major premise. Alternative and disjunctive propositions, however, can have more than two component propositions. Hypothetical syllogisms with multimembered majors are evidently more difficult to evaluate than those with only a two-membered major. Nevertheless, the rules we have been discussing apply proportionally to them if properly modified. Here we will see in what these modifications consist.

The Alternative Inclusive Syllogism

The alternative inclusive proposition states that at least one of several component propositions must be true in the light of the others. Any or all of the others may be true or false. Thus, if any element of a multi-membered alternative inclusive syllogism is affirmed, the others are collectively and singly in doubt. Nothing at all follows in the first figure. But, if one of the elements is denied, a conclusion follows which asserts, in another alternative inclusive proposition, the truth of at least one of the others, e.g.:

$a \vee b \vee c \vee d$
But $-a$
Therefore, $b \vee c \vee d$

When we deny several alternatives we must conclude to an inclusive alternative which asserts the truth of at least one of those remaining.

If but one remains, a categorical proposition follows in which this one is affirmed, e.g.:

1. $a \lor b \lor c \lor d$
   But $-a \cdot -b$
   Therefore, $c \lor d$
2. $a \lor b \lor c \lor d$
   But $-a \cdot -b \cdot -c$
   Therefore, d

The Exclusive Alternative Syllogism

The exclusive alternative proposition asserts hypothetically that one of a number of component propositions is true while the others cannot be true. Thus, in a multimembered type of exclusive alternative syllogism, if any element of the major is affirmed, each of the others must be denied, e.g.:

$a \land b \land c \land d$
But a
Therefore, $-b \cdot -c \cdot -d$

If one element is denied, one, and only one, of the remaining is true, and this is asserted in an exclusive alternative proposition composed of the remaining elements, e.g.:

$a \land b \land c \land d$
But $-a$
Therefore, $b \land c \land d$

If several elements are denied, there are again two possibilities. If but one remains, this must follow as true. If there are several remaining, an exclusive alternative, which asserts that one and only one of these must be true, follows as a conclusion, e.g.:

1. $a \land b \land c \land d$
   But $-a \cdot -b \cdot -c$
   Therefore, d
2. $a \land b \land c \land d$
   But $-a \cdot -b$
   Therefore, $c \land d$

The Disjunctive Syllogism

The disjunctive proposition is a hypothetical assertion that at least one of a number of component propositions is false. Thus, if one of the elements of a multimembered disjunctive proposition is affirmed, a disjunctive conclusion which asserts that one of the others must be false follows, e.g.:

$$n - (a \cdot b \cdot c \cdot d)$$
But a
Therefore, $n - (b \cdot c \cdot d)$

If several are affirmed respectively as true, there are two possibilities. If only one remains, it must be false. If several remain, then a disjunctive proposition which asserts that at least one is false follows as a conclusion, e.g.:

1. $n - (a \cdot b \cdot c \cdot d)$
 But $a \cdot b \cdot c$
 Therefore, $-d$
2. $n - (a \cdot b \cdot c \cdot d)$
 But $a \cdot b$
 Therefore, $n - (c \cdot d)$

XI. Major Premises With Compound Components

We noted earlier that the less complex compound propositions involve categorical elements and that the more complex involve component propositions which are themselves compound: $a \rightarrow b$ is an example of a relatively incomplex compound proposition; $(a \cdot b) \rightarrow (c \wedge d)$ is an example of a more complex compound proposition. The first is a simple conditional with categorical components. The second is a simple conditional whose antecedent is a copulative proposition and whose consequent is an exclusive alternative proposition. $a \rightarrow b$ can stand symbolically for almost all the simple conditionals we have used as examples so far, for in most cases we have limited ourselves to conditionals with categorical components. $(a \cdot b) \rightarrow (c \wedge d)$ could be used symbolically to stand for a proposition such as *If the Communists are militarily powerful and the free world is militarily unprepared, then either the free world will fall to the Communists or God will intervene on the side of the free world, but not both.*

We have seen that conditional, alternative, and disjunctive propositions with categorical components can serve as major premises in hypothetical syllogisms. Similarly, hypothetical propositions with compound components can be used as majors in hypothetical argumentation. When this is the case the minor premise itself will ordinarily be a compound proposition affirming or denying a compound element of the hypothetical major, e.g.:

$$(a \cdot b) \rightarrow (c \wedge d)$$
$$\text{But } -(c \wedge d)$$
$$\text{Therefore, } -(a \cdot b)$$

This is a simple conditional syllogism. The antecedent of its major premise is a copulative proposition and its consequent is an exclusive alternative proposition. Since the minor denies the consequent of the major, the conclusion must deny its antecedent. To illustrate how complex the situation can become with a hypothetical major premise as complex as this one, let us note various additional ways in which the consequent might be denied. If we refer to the truth table for hypotheticals, which cannot be used to confirm hypotheticals taken strictly but sometimes can indicate disconfirmation, we note that $c \cdot d$ and $-c \cdot -d$ both imply $-(c \wedge d)$. Hence, there are at least three other minors which can yield the same conclusion as $-(c \wedge d)$. These are $c \cdot d$, $-c \cdot -d$, and $(c \cdot d) \wedge (-c \cdot -d)$. Nor are these the only ones. $c \wedge d \equiv -c \leftrightarrow d$, and $(-c \leftrightarrow d) \rightarrow (-c \rightarrow d)$. Thus, $-(c \wedge d) \equiv -(-c \leftrightarrow d)$, and $-(-c \rightarrow d) \rightarrow -(-c \leftrightarrow d)$. Hence, both $-(-c \leftrightarrow d)$ and $-(-c \rightarrow d)$ can also be used as minor premises to yield the same conclusion as $-(c \wedge d)$.

There is no need for us to venture further into the possibilities of increasing complexity. We have seen enough to know that hypothetical propositions, and consequently hypothetical syllogisms, can become extremely complex. However, so long as the basic nature of the hypotheticals is understood and the fundamental rules for hypothetical discourse accurately enforced, there is no reason why hypothetical argumentation should ever become unreasonable.

Exercise XVII

1. Define: hypothetical proposition, hypothetical syllogism, simple conditional syllogism, reciprocal conditional syllogism, inclusive alternative syllogism, exclusive alternative syllogism, disjunctive syllogism.
2. What is the distinguishing feature of the hypothetical proposition? How many types of hypothetical proposition are there? How do they differ from one another?
3. What is the rationale, generally speaking, of hypothetical argumentation?
4. How are the figures and the moods of the hypothetical syllogism determined?
5. Why is it legitimate to conclude from the affirmation of the consequent in a reciprocal conditional syllogism, but not in a simple conditional syllogism?

THE HYPOTHETICAL SYLLOGISM 241

6. Why is it illegitimate to proceed from the contrary of the consequent in a simple conditional syllogism to the contrary of its antecedent, when it is legitimate to proceed from the contradiction of the consequent to the contradiction of the antecedent?

7. What type of hypothetical syllogism can be considered to have a categorical equivalent? Illustrate.

8. In what sense is the exclusive alternative syllogism a composite of corresponding inclusive alternative and disjunctive syllogisms?

9. Classify each of the following hypothetical syllogisms according to type and figure. Check each one for validity.
   9.1. If a woman is beautiful, she will have many admirers; but Susan has many admirers; therefore, Susan must be beautiful.
   9.2. Either a plane figure has three angles equal to two right angles or it is not a triangle; but no rectangle has three angles equal to two right angles; therefore, no rectangle is a triangle.
   9.3. Unless some men are students, no men are teachers; but some men are not teachers; therefore, some men are not students.
   9.4. If *color* is the specific difference of *redness*, then not every essential predicate is categorically substance; but *color* is not the specific difference of *redness*; therefore, every essential predicate is categorically substance.
   9.5. Either a liquid boils at 100 degrees centigrade or it is not water; but this liquid boils at 100 degrees centigrade; therefore, this liquid is water.
   9.6. Only if a man is intelligent is he a good leader; but John is intelligent; therefore, John is a good leader.
   9.7. No universal is both the genus and the difference of the same inferior; but *biped* is not the genus of the inferior *man*; therefore, *biped* is the difference of *man*.
   9.8. If and only if a being is alive, it has a soul; but every maple tree is alive; therefore, every maple tree has a soul.
   9.9. Either the principle of contradiction holds or no propositions are certain; but some propositions are certain; therefore, the principle of contradiction holds.
   9.10. If the citizens of a country are generally of low moral caliber, that country is in danger of disintegration from within; but the citizens of Ireland are not generally of low moral caliber; therefore, Ireland is not in danger of disintegration from within.

10. Once again classify each of the following hypothetical syllogisms according to type and figure, and then check each one for validity:
    10.1. $a \lor -b$
          But b
          Therefore, $-a$
    10.2. $-a \leftrightarrow b$
          But a
          Therefore, b

10.3.   n — (a · —b)
        But a
        Therefore, —b
10.4.   a∨b∨c
        But —a · —b
        Therefore, c
10.5.   a∧—b
        But b
        Therefore, a
10.6.   a∧—b∧—c
        But —a · b
        Therefore, c
10.7.   —a↔b
        But a
        Therefore, b
10.8.   n — (—a · —b)
        But —b
        Therefore, a
10.9.   n — (a · b · c)
        But a
        Therefore, n — (b · c)
10.10. —a∧b
        But —a
        Therefore, —b
10.11. {[(a · b) → c] · —(a · b)} → —c
10.12. [(—a↔b) · —b] → a
10.13. {[(a∧b) → c] · —c} → —(a∧b)
10.14. [(a∨b∨—c∨—d) · (c · d)] → a · b
10.15. [n —(—a · —b) · a] → —b

11. Practice the reduction of alternative and disjunctive syllogisms to the conditional by reducing the alternatives and disjunctives in the preceding two questions to equivalent conditionals.

CHAPTER XVIII

Complex Syllogistic Patterns

I. Complex Syllogisms

Thus far our study of argument has been limited to a generally uncomplicated form of syllogism. We have, to be sure, studied many different syllogistic patterns, but each has involved only a major premise and a minor premise, together yielding a conclusion. However, we would be seriously misled were we to conclude that every syllogism must be exclusively of one type, with one figure and one mood, and that each must be composed of one conclusion following from two and only two premises. The student will find that many syllogisms are relatively complex, representing a more or less complicated combination of several of the syllogisms of the type we have so far studied. It is true that the logical force of these complex syllogisms stems from the simpler syllogisms from which they are constructed. Thus, we have already investigated all that we need fundamentally in order to evaluate complex syllogisms. Nevertheless, it is helpful to consider at least some of the possible complex syllogistic patterns. Otherwise they can prove to be stumbling blocks for a student when he meets them in other courses. Some complex syllogistic patterns are rarely encountered. Others are almost commonplace. We shall consider the most significant of the latter in this chapter.

II. The Abbreviated Syllogism

Before taking up the most significant types of complex syllogism, let us consider what is spoken of as the abbreviated syllogism. Whatever an abbreviated syllogism is, certainly it is not one of the types of complex syllogism spoken about in the preceding section. There we indicated that by a complex syllogism we refer to a more or less complicated syllogistic pattern involving a combination of fundamentally simple or incomplex categorical and/or hypothetical syllogisms. Thus, the types of complex syllogism are always expanded — never abbreviated — syllogisms.

Some logicians speak of the abbreviated syllogism as a special type of syllogism. They call it an enthymeme and define it as a syllogism in which one proposition is not expressed but nevertheless understood. This calls for some comment. First of all, the term "syllogism" can be used for both the mental deductive argumentation and its oral or written expression. It is possible to omit the oral or written expression of a proposition which is nonetheless understood. But it is impossible to understand a proposition and at the same time to fail mentally to express it, for it is understood precisely in so far as it is mentally expressed. Thus, we can legitimately speak of an abbreviated syllogism only in reference to the oral or written expression of a logical entity. When we talk about the special types of syllogism in this chapter we are concerned with more than verbal expression.

The practice of some logicians to speak of the abbreviated syllogism as a special type of syllogism and to call it an enthymeme calls for a second comment. The term "enthymeme" is borrowed from Aristotle, who used it for the syllogism which is characteristic of rhetoric. It is not our purpose here to discuss the typical characteristics of the rhetorical syllogism. It is enough to say that the difference between the syllogism of the rhetorician and any other syllogism is not a difference in form, but rather a difference in matter. The rhetorician argues from premises which express probabilities and signs, but unless he argues from at least two premises he can come to no conclusion. It is true that in orally expressing himself the rhetorician frequently uses an abbreviated form. More often than not he may leave one or other of his premises unexpressed (orally, that is, but not mentally). It is because of this that the term "enthymeme" has become associated with the notion of the abbreviated syllogism, and through this association it has come to be used by some generally to refer to all abbreviated syllogisms. We should note that to use the term in this way is not to use it as Aristotle used it. What is more important, we must note that it is meaningless to speak at all of an abbreviated syllogism except in the context simply of the grammatical expression of a logical expression which defies mental abbreviation.

Although there is no such thing as an abbreviated syllogism from the strictly logical point of view, we must be prepared to meet abbreviated grammatical expressions of syllogisms, and we should be able to use them ourselves. Usually a premise is the element not expressed in so-called abbreviated syllogisms, although at times the element not expressed is the conclusion. In any event these syllogisms can be evaluated only after the unexpressed premise or conclusion is supplied. Then, depending on the type of syllogism involved, the resulting syllogism must be evaluated in

COMPLEX SYLLOGISTIC PATTERNS   245

the light of the appropriate rules for syllogistic procedure. If the syllogism, with the unexpressed premise or conclusion supplied, measures up to the rules appropriate to it, it is valid. If it fails to measure up, it is invalid.

To argue that no plant is able to suffer because every plant is nonsentient is to argue validly, so long as the unexpressed premise which is understood is *Nothing which is nonsentient can suffer*. Fully expressed, this is a valid first figure categorical syllogism concluding directly.

>Nothing which is nonsentient can suffer.
>But every plant is nonsentient.
>Therefore, no plant can suffer.
>(*Valid.*)

However, to argue that the number seven is odd because it is a whole number is invalid, if the unexpressed premise is *If a number is odd, it is a whole number*. Fully expressed, this is a first figure simple conditional syllogism which is invalid because it attempts to go from the affirmation of the consequent to the affirmation of the antecedent.

>If a number is odd, it is a whole number.
>But the number seven is a whole number.
>Therefore, the number seven is an odd number.
>(*Invalid:* One cannot affirm the consequent in order to affirm the antecedent.)

"Every creature is dependent on the Creator for existence and some creatures are men" is a valid categorical syllogism concluding directly in the third figure, if the unexpressed conclusion is understood to be *Some men are dependent on the Creator for existence*. It is an invalid syllogism with an illicit minor term if the conclusion which is understood is *Every man is dependent on the Creator for existence*.

1. Every creature is dependent on the Creator for existence.
   But some creatures are men.
   Therefore, some men are dependent on the Creator for existence.
   (*Valid.*)
2. Every creature is dependent on the Creator for existence.
   But some creatures are men.
   Therefore, every man is dependent on the Creator for existence.
   (*Invalid:* The syllogism has an illicit minor term.)

III. The Syllogism With a Justification for Its Premises

Perhaps the expanded syllogism most frequently used is one (either categorical or hypothetical) which adds a justification to one or both of its premises. This occurs whenever a premise is expressed precisely as a conclusion or as the main component in a causal proposition. To illustrate:

1. No polygon is a circle, because every polygon is rectilinear, and no circle is rectilinear.
   But every square is a polygon.
   Therefore, no square is a circle.
2. All concepts identical in comprehension are identical in extension.
   But *man* and *rational animal* are identical in comprehension since the latter is the essential definition of the former.
   Therefore, *man* and *rational animal* are identical in extension.
3. If any A proposition has the definition of its subject for its predicate, that A proposition can be effectively simply converted, because the definition is convertible with the defined.
   But *Every man is a rational animal* is an A proposition with a predicate which is the definition of its subject, since *animal* is the proximate genus of *man* and *rational* is the specific difference of *man*.
   Therefore, *Every man is a rational animal* can be effectively simply converted.

Logicians have called this type of expanded syllogism an epicheirema, although once again we should note that this is not the way in which Aristotle used this term. Aristotle used the term to refer to a dialectical syllogism proceeding from probable premises. However, since this is a legitimate type of syllogism, we shall speak of it as an epicheirema, for want of a better term, even though this is not the way in which Aristotle used this term. Thus understood, the epicheirema is basically a simple syllogism involving a major premise and a minor premise, with some justification expressed for one or both of its premises. The epicheirema can be checked for validity by separately checking the sequence of its basic syllogistic pattern on the one hand and the sequence between its premises and their antecedents on the other. The epicheirema is valid only if there is no invalid sequence in its total structure.

The following syllogism (1) is a valid epicheirema. It has a basic syllogistic structure (2) which is a valid first figure categorical syllogism concluding directly. It has a major premise (3) which is the conclusion of a valid second figure categorical syllogism concluding indirectly.

1. No polygon is a circle, because every polygon is rectilinear, and no circle is rectilinear.
   But every square is a polygon.

COMPLEX SYLLOGISTIC PATTERNS 247

Therefore, no square is a circle.
(*Valid* epicheirema.)
2. No polygon is a circle.
But every square is a polygon.
Therefore, no square is a circle.
(*Valid* first figure categorical syllogism concluding directly.)
3. Every polygon is rectilinear.
But no circle is rectilinear.
Therefore, no polygon is a circle.

(*Valid* second figure categorical syllogism concluding indirectly.)

The following is an epicheirema with a valid basic categorically syllogistic pattern in the first figure concluding directly. It is invalid because it has a major premise which is the conclusion of an invalid second figure categorical syllogism with an unextended middle term.

Every man is capable of speech, because everyone capable of speech is rational, and every man is rational.
But John is a man.
Therefore, John is capable of speech.

Our final example is an invalid epicheirema. It is basically an invalid categorical syllogism suffering from an unextended middle term despite a minor premise which is a legitimate causal proposition.

Every animal is sentient.
But some organisms are sentient, since everything sentient is an organism.
Therefore, some organisms are animal.

IV. The Polysyllogism

The polysyllogism is a clear-cut case of an expanded syllogism. It is compounded from several simple syllogisms in such fashion that the conclusion of one simple syllogism is used as the major premise for the next until finally the conclusion of the polysyllogism taken as a whole is reached. The component syllogisms in the polysyllogism can be either categorical or hypothetical. To illustrate:

1. No being-of-the-reason is categorized.
But every logical relation is a being of the reason.
Therefore, no logical relation is categorized.
But every predicable as such is a logical relation.
Therefore, no predicable as such is categorized.

2. If seven is a whole number, seven is either odd or even.
      But seven is a whole number.
      Therefore, seven is either odd or even.
      But seven is not even.
      Therefore, seven is odd.

To check the validity of a polysyllogism simply reduce the polysyllogism to the simple syllogisms out of which it is constructed, and then check each component simple syllogism against the appropriate rules. If any of the simple syllogisms is invalid, the polysyllogism as a whole is invalid. If all of the component syllogisms are valid, the polysyllogism is valid. The preceding arguments are valid polysyllogisms. The first is resolved into two valid first figure categorical syllogisms concluding directly. The second is resolved into a valid conditional syllogism in the first figure and a valid alternative syllogism in the second figure. The following arguments are invalid polysyllogisms:

   1. If every animal is mortal, then every animal is a substance.
      But every animal is a substance.
      Therefore, every animal is mortal.
      But every man is an animal.
      Therefore, every man is mortal.
   2. Every free agent is morally responsible for his actions.
      But every man is a free agent.
      Therefore, every man is morally responsible for his actions.
      But Tom is morally responsible for his actions.
      Therefore, Tom is a man.

The first of these invalid polysyllogisms includes two simple syllogisms: the first is an invalid simple conditional syllogism in the first figure, and the second is a valid categorical syllogism in the first figure concluding directly. The second polysyllogism includes, first of all, a valid categorical syllogism in the first figure concluding directly and, second, an invalid categorical syllogism in the second figure concluding directly.

V. The Sorites

The sorites is a special type of extended syllogism which is ordinarily an expanded categorical syllogism with more than three terms and more than two premises. It is, in fact, nothing but a telescoped form of polysyllogism. There are two ways of telescoping a polysyllogism so that it becomes a sorites: one results in what is known as the Aristotelian sorites; the other, the Goclenian sorites.[1] The Aristotelian sorites is constructed of a number of propositions so arranged that the predicate of one becomes the subject of the next, until finally the conclusion is reached in which the predicate of the last premise is either affirmed of or denied of the subject of the first premise. In the Goclenian sorites the subject of one proposition becomes the predicate of the next, until the conclusion is reached in which the predicate of the first premise is affirmed or denied of the subject of the last premise. The Aristotelian sorites has a series of middle terms with increasing extension, while the Goclenian sorites has a series of middle terms with decreasing extension. The following argumentations are respectively examples of the Aristotelian and Goclenian sorites:

1. Every man is an animal.
   But every animal is an organism.
   And every organism is a body.
   And every body is a substance.
   Therefore, every man is a substance.
   (*Valid* Aristotelian sorites.)

2. Every body is a substance.
   But every organism is a body.
   And every animal is an organism.
   And every man is an animal.
   Therefore, every man is a substance.
   (*Valid* Goclenian sorites.)

The validity of a sorites, like that of a polysyllogism, depends upon the validity of each of the simple syllogisms out of which it is constructed. If any of them is invalid, the sorites itself is invalid. However, unlike the polysyllogism, the sorites has a special set of rules against which it can be measured. These rules are simply restatements of the rules for categorical syllogism contracted precisely to the special case of the sorites. Clearly each of them holds because its violation ultimately violates one of the general rules for categorical syllogism. There are three rules for a valid sorites:

RULE 1. *No premise may be particular except the one which contains the subject of the conclusion (i.e., the first premise in the Aristotelian sorites and the last premise in the Goclenian sorites).*

---

[1] Aristotle did not use the term "sorites," but he did suggest the form of syllogism known as the Aristotelian sorites. The so-called Goclenian sorites was proposed by the logician Goclenius, a sixteenth-century commentator on the logical works of Aristotle.

The following Aristotelian sorites breaks the rule:

> Every A is B.
> But every B is C.
> And some C is D.
> Therefore, some A is D.

(This is invalid because C, which is used as a middle term, is never universal. If the second premise is made negative in order to make C universal, then the sorites involves an affirmative conclusion coming from a negative premise. If the conclusion is made negative to escape this, then D in the conclusion is more universal than D in the last premise. There is no way to escape the charge of invalidity so long as any premise except the one which contains the subject of the conclusion is particular.)

RULE 2. *No premise may be negative except the one which contains the predicate of the conclusion (i.e., the last premise in the Aristotelian sorites and the first premise in the Goclenian sorites).*

The following Goclenian sorites breaks this rule:

> Every A is B.
> But no C is A.
> And every D is C.
> Therefore, no D is B.

(This is invalid because B is universal in the conclusion and only particular in the first premise. If the first premise is made negative to escape this, there are two negative premises. If the conclusion is made affirmative to escape the same thing, there is a negative premise and an affirmative conclusion. In any event, there is some defect in the argumentation so long as any premise except the one containing the predicate of the conclusion is negative.)

RULE 3. *A particular premise necessitates a particular conclusion, and a negative premise necessitates a negative conclusion.* This is true, of course, precisely in virtue of the general rules for the categorical syllogism which demand precisely the same thing.

The following sorites break this third rule:

1. Some A is B.
   But every B is C.
   And every C is D.
   Therefore, every A is D.

(This is invalid precisely because A is universal in the conclusion and only particular in the first premise.)

2. Every A is B.
   But every C is A.
   And no D is C.
   Therefore, every D is B.

(This is invalid precisely because no affirmative conclusion is possible following a negative premise.)

Although a sorites is ordinarily an expanded form of categorical argumentation, it is possible to construct a conditional sorites. In the conditional sorites several premises are so constructed that the consequent of one becomes the antecedent of the next. In accord with the rules for conditional argumentation the conclusion which follows from these premises must combine the antecedent of the first premise with the consequent of the last. To illustrate:

> If a plane figure is a square, it is a rectangle.
> But, if it is a rectangle, it is a parallelogram.
> And, if it is a parallelogram, it is four-sided.
> Therefore, if a plane figure is a square, it is four-sided.

Because the denial of the consequent in a conditional proposition implies the denial of its antecedent, the preceding conditional sorites can be restated in the following fashion. This reveals a second form of conditional sorites in which the antecedent of one premise becomes the consequent of the next until a conclusion is reached in which the consequent of the first is joined to the antecedent of the last.

> If a figure is not a rectangle, it cannot be a square.
> And, if a figure is not a parallelogram, it cannot be a rectangle.
> And, if a figure is not a four-sided figure, it cannot be a parallelogram.
> Therefore, if a figure is not a four-sided figure, it cannot be a square.

VI. The Dilemma

The dilemma is a complicated form of syllogistic argument combining conditional and alternative propositions in an antecedent which yields either a categorical or an alternative conclusion. The major premise in any dilemma is a complex copulative proposition composed of simple conditionals. The minor premise is an alternative proposition which alternately affirms the antecedents or denies the consequents of the conditionals in the major. There are four different forms that a dilemma can take. These constitute the simple constructive dilemma, the complex constructive dilemma, the simple destructive dilemma, and the complex destructive dilemma. A simple dilemma differs from a complex dilemma in that the conditional propositions in its major premise have either a common antecedent or a common consequent, whereas the conditional propositions in the major of the complex dilemma have both different antecedents and different consequents. The constructive dilemma differs from the destructive in that the minor premise in the former alternately affirms the antecedents of the conditionals in the major, while the minor premise in the latter alternately denies the consequents of the conditionals in the major premise.

The Simple Constructive Dilemma

In the simple constructive dilemma each of the conditionals in the copulative major premise has the same consequent. Their antecedents are the component propositions in the alternative minor. The major and minor premises together yield as a conclusion the consequent common to the component conditionals of the major premise.

For example:

$(a \rightarrow c) \cdot (b \rightarrow c)$
But $a \lor b$
Therefore, c

The Simple Destructive Dilemma

In the simple destructive dilemma the conditional major premise expresses two consequents following from a common antecedent. The minor premise alternately denies each of these consequents. The denial of the antecedent common to the two consequents follows as a conclusion.

To illustrate:

$a \rightarrow (b \cdot c)$
But $-b \lor -c$
Therefore, $-a$

The Complex Constructive Dilemma

In the complex constructive dilemma the copulative major is composed of conditionals with different antecedents, each implying different consequents. The alternative minor premise alternately affirms the antecedents. A conclusion follows which alternately affirms the consequents.

As an example:
$(a \rightarrow c) \cdot (b \rightarrow d)$
But $a \lor b$
Therefore, $c \lor d$

The Complex Destructive Dilemma

The major premise of the complex destructive dilemma is composed of conditionals with respectively different antecedents and consequents. The minor alternately denies each of the consequents. A conclusion follows which alternately denies each of the antecedents.

For example:
$(a \rightarrow c) \cdot (b \rightarrow d)$
But $-c \lor -d$
Therefore, $-a \lor -b$

The dilemma is a valid form of syllogistic argumentation. It is composed of conditional and alternative propositions, and it is constructed strictly in accord with the rules for the conditional and alternative syllogisms. So long as a dilemma is proposed in one of its four distinctive forms, and so long as the hypothetical propositions out of which it is formed are worthwhile, the dilemma is a legitimate complex syllogism. Defective hypothetical components will, of course, render a dilemma itself defective. Logicians speak of destroying a dilemma either by slipping between its horns, or by taking it by its horns. To slip between the horns of a dilemma is to show that its minor premise is illegitimate by pointing out an additional component for the alternative premise. To take the dilemma by the horns is to show that one or other of the consequents in the conditionals comprising the major premise is not implied by its supposed antecedent.

VII. Combinations of Complex Syllogisms

The complex syllogisms we have investigated in this chapter are combinations of the simple syllogisms previously considered. There is no reason why these complex syllogisms cannot themselves be combined into still more complicated complex patterns. So long as each of the component syllogisms in the complex syllogism which results measures up to its own set of rules, the complex syllogism as a whole is valid. Consider the following example of a complex syllogism which finally yields the conclusion *Khrushchev cannot be trusted:*

> If a Communist is a good Communist, he cannot be trusted, because on principle he is committed to the thesis that the end justifies the means. If a Communist is a bad Communist, he cannot be trusted, because he is not true even to his own principles. But a Communist must be either a good Communist or a bad Communist. Hence, no Communist can be trusted. Thus, since Khrushchev is a Communist, Khrushchev cannot be trusted.

This is an example of a valid polysyllogism constructed from a simple constructive dilemma and a first figure categorical syllogism concluding directly. The conclusion of the dilemma is the major premise of the categorical syllogism. The dilemma itself is an example of an epicheirema, since there is a justification added to each of the conditionals comprising the major premise of the dilemma.

Exercise XVIII (A)

1. Define: complex syllogism, abbreviated syllogism, epicheirema, polysyllogism, sorites, dilemma.
2. What is the logical status of the abbreviated syllogism?
3. What are the rules for the validity of an epicheirema?
4. What are the rules for the validity of a polysyllogism?
5. What is the difference between the Aristotelian and the Goclenian sorites? Defend the rules for a valid sorites.
6. What is the general plan of a dilemma? What are the different forms of the dilemma? When is a dilemma a legitimate form of argumentation?
7. Suggest examples of both valid and invalid epicheiremas, polysyllogisms, sorites, dilemmas, and any combinations of these. Explain the defect in the invalid examples.
8. Identify the special type of syllogism exemplified in each of the following. Evaluate each as far as validity is concerned, and explain what is wrong with each invalid argumentation:
   - 8.1. Every A is B, because some B is A.
     But some A is C, because every C is A.
     Therefore, some C is B.
   - 8.2. No A is B.
     But every B is C.
     And every C is D.
     And every D is E.
     And some E is G.
     Therefore, some A is not G.

COMPLEX SYLLOGISTIC PATTERNS   255

8.3. Every A is B.
     But no C is A.
     Therefore, no C is B.
     But some C is D.
     Therefore, some D is not B.

8.4. If a proposition is self-evident, it is immediate, since it can be seen apart from premises.
     But The White Mountains are beautiful is immediate, since it can be seen apart from premises.
     Therefore, The White Mountains are beautiful is self-evident.

8.5. $(a \rightarrow -c) \cdot (-b \rightarrow d)$
     But $-c \lor d$
     Therefore, $a \lor -b$

8.6. Whatever is free is intelligent.
     But whatever is intelligent is spiritual.
     And whatever is spiritual is incorruptible.
     Therefore, whatever is incorruptible is free.

8.7. $-a \lor b$, since $a \rightarrow b$
     But $-b$, since $-(b \cdot c)$
     Therefore, $-a$

8.8. Every polygon is a rectilinear plane figure.
     But every parallelogram is a polygon.
     And every rectangle is a parallelogram.
     And no circle is a rectangle.
     Therefore, no circle is a rectilinear plane figure.

8.9. $(a \rightarrow c) \cdot (-b \rightarrow -d)$
     But $a \lor -b$, since $a \rightarrow b$
     Therefore, $c \lor -d$, since $-c \cdot d$
     But d
     Therefore, c

8.10. $(a \rightarrow b) \rightarrow (c \lor d)$
      But $a \rightarrow b$, because $e \rightarrow b$, and $e \equiv c$
      Therefore, $c \lor d$
      But $-c$, because $f \rightarrow c$, and $-f$
      Therefore, d

Exercise XVIII (B)

This is a general exercise in formal logic for all types of syllogistic argumentation. You are to identify the type of syllogism which is attempted in each of the following argumentations. Wherever possible indicate its figure and its mood. For this you may find it helpful to restate the syllogisms in strictly logical form wherever this is not already the case. You may find it helpful also to restate each argumentation symbolically. Check each syllogism for validity, and explain what is wrong with each invalid syllogism. Supplement this exercise by examining this textbook, as well as your other textbooks and reference books, for attempts at syllogistic argumentation, classifying and evaluating each in the manner suggested above for the following examples of argumentation:

1. Unless a predicable is necessary, it is not essential. Thus, the predicable property is essential, because it is necessary.
2. One must surely accept what one can plainly see. Hence, these words must be accepted, since they can be plainly seen.
3. Acts of worship are not contrary to nature, since no act is both contrary to nature and pleasing to God, and acts of worship are pleasing to God.
4. Since one cannot be both an old-guard Republican and politically liberal, the politician in question must be an old-guard Republican, because he is certainly not politically liberal.
5. A coward cannot help but hurt himself; since cowards lie, and liars cheat, and cheaters hurt themselves.
6. Either a composite expression is a perfect composite expression or it is not a proposition, for every proposition is a perfect composite expression. Therefore, no definition is a proposition.
7. Either a composite expression is a proposition or it is not a perfect composite expression, for every proposition is a perfect composite expression. Thus, no definition is a proposition, because no definition is a perfect composite expression.
8. Hell strikes fear in our hearts because it is a frightening word, and frightening words strike fear in our hearts.
9. The human soul cannot be corrupted, because it is not composed of entitative parts, and every corruptible substance is composed of entitative parts.
10. Some polygon is rectangular, and every polygon is rectilinear; hence, every rectangle is rectilinear.
11. Only animals are sentient. Thus, man is sentient, because man is animal.
12. If 20 is higher than 15, it is higher than 10; but 20 is higher than 10; therefore, 20 is higher than 15.
13. Either the human will is completely determined, or it is completely undetermined, or it is partially determined. If it is completely determined, we cannot be held accountable for our choices. If it is completely undetermined, we cannot choose at all. Since we do choose and are held accountable for our choices, the human will must be partially determined.
14. Some Romans are good singers, since all Romans are Italian, and many Italians are good singers.
15. No circles are rectilinear, because all rectangles are rectilinear, and no circles are rectangles.
16. If and only if a figure has interior angles equal to 180 degrees, it is a triangle. Thus, since every isosceles figure is a triangle, every isosceles figure has interior angles equal to 180 degrees. Accordingly, since no scalene triangle is isosceles, no scalene triangle has interior angles equal to 180 degrees. This, however, implies that no scalene triangle is a triangle.

17. John cannot be a Marine, because his military record is anything but enviable, and the Marines have an enviable military record.
18. No one can be simultaneously a father and a son. Thus, since Tom is a son, Tom is not a father.
19. Seven must be an odd number, since it is not even.
20. Since a Fascist government cannot be Communist, and a Communist government cannot be Fascist, then a non-Communist government must be Fascist, and a non-Fascist government must be Communist. Thus, every government must be either Fascist or Communist.
21. Immorality is the greatest of all evils. Thus, cheating on an exam is the greatest of all evils, since cheating on an exam is an immorality.
22. Our government is no stronger than our chief executive; hence, a strong president insures a strong government.
23. An American either is loyal to his country, or he is not loyal to his country. If he is a Communist, he is not loyal to his country. But he must either be a Communist or not be a Communist. Hence, either an American is a Communist or he is a loyal American.
24. No immediate proposition is proved, because every self-evident proposition is immediate, and no self-evident proposition is proved.
25. Every science is a quality, for every habit is a quality, and every intellectual virtue is a habit, and every science is an intellectual virtue.
26. If a man is psychotic, he is not normal. Thus, John is psychotic, because he is not normal; while Peter is normal, since he is not psychotic.
27. Running is a transitive operation, because it is categorized as action, and every transitive operation is categorized as action.
28. To make political broadcasts sympathetic to an enemy is to collaborate, and to collaborate is to do something to be ashamed of; thus, many of our prisoners of war from the Korean War have nothing to be ashamed of, since many of them did not make political broadcasts sympathetic to the enemy.
29. Since a liberal favors foreign aid, a politician who does not favor foreign aid is not a liberal, and a politician who does favor foreign aid is a liberal.
30. All animals are sentient, since no plants are sentient, and no animals are plants.

CHAPTER XIX

Demonstrative and Dialectical Discourse

I. The Syllogism Materially Considered

Thus far we have considered the syllogism solely from the point of view of formal logic. We have investigated exclusively what it takes to make a syllogism valid. But validity does not guarantee truth. Consequently, it is hardly enough to master merely the rules of valid argument. Logic looks beyond itself to *totally* sound discourse in the other disciplines. The logic student must learn not only what makes a conclusion follow from its premises, but what makes it follow determinately either as certainly true or as probably true. Thus, it remains for us to investigate the logical properties belonging to syllogisms from the point of view of their matter. These are the properties which determine, in the general way proper to logic, the truth, certainty, and explanatory force of the syllogism. Several syllogisms identical in form can have conclusions with significantly different force as far as truth is concerned. One may involve logical properties guaranteeing at best a probably true conclusion, which admits of the possibility of its opposite. A second may involve logical properties which guarantee a factually true conclusion, which most certainly is true but could, given another factual situation, not-be true. A third may involve logical properties which guarantee a necessarily true conclusion, which is true precisely because it cannot not-be true. The logical properties which determine the force of a conclusion in reference to truth and certitude are the subject matter of this chapter.

II. Science and Demonstration

The word "science" is used in many ways. It is used most commonly today to refer to the experimental or laboratory disciplines, such as physics, chemistry, and biology, which are perhaps most aptly described as empiriological. This is an acceptable usage of the term "science," but it is not the only one. Aristotle used it in a different way, and his usage has remained respectable although it is much less popular than the one common today. In Aristotle's sense, to know a thing scientifically is to know it in its necessary causes. Scientific knowledge, understood in this sense, is a relatively perfect kind of knowledge, a knowledge in which the intellect can rest because it is both certain and causally explained. It is a kind of knowledge achieved, unfortunately, only with great difficulty and all too rarely. Yet it can be reached, and when it is, it comes as the conclusion to a syllogism which is perfect not only formally but materially as well. A syllogism which can generate a conclusion scientific in the Aristotelian sense — that is, a certain proposition proven through causes — is called a demonstration. Throughout this chapter we shall speak of "science" in Aristotle's sense, and we shall refer to demonstration as the scientific syllogism.

Demonstration itself is an analogous notion. It is realized proportionally in different types of syllogisms *which yield certain knowledge proven through causes* in respectively different ways. We shall distinguish between the types of demonstration very shortly. However, first we must consider the most perfect type of demonstration. Since demonstration is an analogous notion, the others can be understood only by reference to its primary instance. Science, in its strictest sense, is achieved as the conclusion to a syllogism which establishes the necessary connection between a subject and one of its specific properties. It does this in the light of premises which show the necessary connection between this subject and its definition and between this definition and the property in question. The following argument is an example of the strictest form of demonstration:

Every rational animal is capable of speech.
But every man is a rational animal.
Therefore, every man is capable of speech.

This is a categorical syllogism in the first figure concluding directly, with two A premises and an A conclusion. The minor term is a subject of scientific inquiry. The major term is a specific property of this subject. The middle term is simultaneously the necessary and proximate cause of the property of the subject and the essential definition of the subject itself. Since the premises are necessarily true the conclusion follows as necessarily true. Since the premises express the cause of the inherence of the property in the subject the conclusion follows as fully explained. The example may well disappoint the student. It seems all too easy, and perhaps somewhat less than crucial. However, it is probable that the demonstration is not yet fully appreciated by the student. It is one thing to admit *that* the premises are necessary and explanatory and *that* the conclusion is therefore necessary and explained. It is another to see the necessary and explanatory character of the premises. This involves keen intellectual insight, which comes perhaps only as the result of lengthy and laborious intellectual effort. However, once the premises are fully appreciated for what they are in themselves, it is relatively simple to view them as premises and to appreciate the scientific status of the conclusion generated from them. For this reason the demonstration itself may seem to be anticlimactic. We shall say more, in this chapter and in the next, about the premises in the demonstration and the effort involved in achieving them.

III. Prescientific Knowledge

We have seen that scientific knowledge, in Aristotle's sense of the term, is achieved as a conclusion in a demonstrative syllogism. Because it is a conclusion, the scientific proposition presupposes prior knowledge. In this section we shall consider what must be known before it is possible to demonstrate a conclusion. First of all, the terms of any proposition must be known before that proposition itself is known. Second, the premises must be known before the conclusion based on them can be known. For demonstration, then, there must be foreknowledge of the subject and predicate of the conclusion of the demonstration, as well as foreknowledge of its premises. This means, for the strictest form of demonstration, that we must have a prior knowledge of its scientific subject, of the property in question, and of the premises of the demonstration.

There are two things we can know about an object. We can know *what* it is, and we can know *that* it is. To know what it is, is to be able to define it. To know that it is, is to know of its existence. The incomplex object of simple apprehension can be known in both ways. We can know what it is either in a nominal or a real definition, and we can know that it is. But the complex object of judgment is different. Propositions cannot be defined, and thus we cannot know *what* they are. But we can know that they are — at least in the way in which propositions are, that is, as *being true*.

With these distinctions in mind, let us consider precisely what must be known before we can demonstrate a conclusion in the strictest form of demonstration:

1. About the *subject* — we must know that the subject exists and we must be able to define it in a real and causal definition. In the strictest form of demonstration we attempt to show that a scientific subject

exists as necessarily possessed of a given property. We show this by establishing a necessary cause-and-effect relationship between the definition of the subject and the property. The proof itself does not bear upon the existence of the subject. This is presupposed. It does bear upon the inherence of the property in the subject. This is shown through a middle term which is the real definition of the subject and the cause of its property. Thus, a real and causal definition of the subject is also presupposed.

2. About the *property* — we must know at least its nominal definition. We do not presuppose knowledge of its existence or its essential definition. A property is categorically an accident; as such, it is essentially defined in terms of its proper subject. To know that an attribute is defined essentially in terms of a given subject is to know that it necessarily inheres in that subject, and that is what is expressed in the scientific conclusion. This need not, and cannot, be presupposed. Nor need the existence of the property be presupposed. Once the premises have established that a property, known only through a nominal definition, actually does flow necessarily from the definition of a subject known to exist, the conclusion must affirm the actual existence of that property in the subject. Thus all that need be presupposed about the property is its nominal definition.

3. About the *premises* — we must know that they are necessarily true.

IV. The Requirements for the Premises of Demonstration

The definition of demonstration as a syllogism productive of science is a definition in terms of final cause. This definition of demonstration through final cause leads us to a definition of demonstration through its matter, namely, a syllogism with premises that are true, primary, immediate, prior to the conclusion, better known than the conclusion, and cause of the conclusion. Unless a syllogism has such premises, it cannot yield a conclusion measuring up to the strict notion of science. Let us see why this is so by considering these characteristics of the premises one by one:

1. The premises must be *true*, for only true premises can *cause* the conclusion *to be true*. False premises allow for, but in no sense ever guarantee, a true conclusion.

2. The premises must be *primary*. This means that they must be principles of demonstration which are themselves not demonstrated (or ultimately reduced to such propositions). The alternative to this is an

absurd infinite regress in premises. In the strict sense a premise is a primary principle of demonstration only if it is proper to its conclusion, that is, only if it is strictly proportioned to the scientific subject. This requires, for example, that whenever a conclusion is about a mathematical subject the middle term in the premises be expressed precisely in a mathematical fashion. A statement out of logic or out of physics would not function *properly* as a *primary* premise in mathematical discourse.

3. The premises must be immediate. This means that the predicate in a premise must belong immediately (that is, not by way of any middle term) to its subject. Again, unless we have this, at least reductively, there is an impossible infinite regress in premises.
4. The premises must be prior to the conclusion because they are principles from which the conclusion proceeds.
5. The premises must be better known than the conclusion because they serve as principles which manifest the conclusion, which in itself lacks the evidence they supply.
6. The premises must be the cause of the conclusion. In one sense this is true of all premises. It is the very nature of argument that the conclusion be virtually in the premises as an effect in a cause. Premises and conclusion are always so related that a knowledge of the premises as premises cause a knowledge of the conclusion. The premises in the strict form of demonstration are related to the conclusion as cause to effect in a very special way. The premises not only cause the conclusion to be known, but they represent the real cause which explains the reality represented in the conclusion.

The conclusion to a strict demonstration is a universal proposition expressing the necessary inherence of a property in a scientific subject. This conclusion can be necessary only in so far as some necessity in the premises is communicated to the conclusion, for nothing is in the conclusion which is not at least virtually in the premises. To say that the premises must be true, primary, and immediate is to insist that one or both of the premises be self-evident. It is not necessary that both premises in a demonstration be self-evident. In the strictest form of demonstration both will be self-evident, but it is possible to demonstrate less strictly with one self-evident and one factually evident proposition. However, at least one premise must be self-evident, at least reductively, or else the conclusion cannot be necessary. This is true because the only absolutely necessary propositions are those which are self-evident or those which follow from strictly self-evident propositions. We have already spoken about self-evident propositions, and we shall discuss them more at length in the following chapter. From the positive point of view a self-evident proposition is a proposition whose truth is evident in the very terms involved. From the negative point of view a self-evident proposition is a universally necessary proposition which involves no middle term between its predicate and its subject.

V. Universality, Perseity, and Convertibility

Three significant logical relations found in the propositions making up the strictest form of demonstration deserve special mention. Not every proposition in every demonstration involves all three. However, a demonstration in which any of these properties is missing is less perfect than one in which they are found. The three relations in question are universality, perseity, and convertibility. Let us consider them one by one:

1. Universality

We have spoken of universality from the formal point of view in previous chapters. In the present reference we add something to this notion. A proposition involves a kind of universality whenever the predicate is said (in any way) of each and every instance of the subject. A proposition involves perfect universality whenever the predicate is said necessarily of each and every instance of the subject. It is this perfect universality which is a property of demonstration.

2. Perseity

Perseity is a logical property belonging to a universal proposition in so far as the predicate is said necessarily of the subject precisely in virtue of what is essential to the subject. The opposite of a *per se* connection between predicate and subject is a *per accidens* connection, in which the predicate is connected with the subject only accidentally. The modes of perseity have been traditionally numbered as four. However, only the first, second, and fourth concern us. The third is a mode of being, not a mode of predication.

A proposition is said to be in the first mode of perseity whenever its predicate is an element in the essential definition of its subject (i.e., either the genus, difference, or essential definition itself). A proposition is in the second mode of perseity whenever its subject is an element in the definition of its predicate. This occurs when the predicate of the proposition is related to the logical subject as an accident to its proper real subject, e.g., as *risible* is related to *man*. A proposition is in the fourth mode of perseity whenever its subject is related to its predicate as the proper (i.e., proximate and precise) and necessary cause of the predicate.[1] Consider the example we have taken of a strict type of demonstration:

> Every rational animal is capable of speech.
> But every man is a rational animal.
> Therefore, every man is capable of speech.

The major premise is in the fourth mode of perseity. The minor premise is in the first mode of perseity. And the conclusion is in the second mode of perseity.

3. Convertibility

A proposition is convertible in this reference whenever its subject and predicate are necessarily coextensive. A convertible proposition is said to be commensurately universal. The predicate belongs to every instance of the subject, and only to this subject. The propositions in our example of the strict type of demonstration are not only universal, but they are commensurately universal as well.

VI. Types of Demonstration

We have concentrated so far on the strictest form of demonstration. There are, in fact, three basically different types of demonstration, with variations possible within each. Each is, in its own way, a demonstration because each generates certain knowledge proved through causes. Demonstration is an analogous notion realized primarily in the kind of demonstration we have discussed so far. We shall call the primary instance of demonstration explanatory demonstration. The other forms of demonstration are less strict instances of demonstration. They are more or less perfect as they more or less closely approach the perfection of the explanatory demonstration.

Demonstration can be divided into types from two different points of view. From one point of view, demonstration is either explanatory demonstration (*propter quid*) or demonstration of the fact (*quia*). Explanatory demonstration not only manifests with certitude the fact that something is necessarily true, but it expresses the proper and adequate reason why it must be true and could not not-be true. Demonstration of the fact simply manifests with certitude the fact that something is necessarily

---

[1] Traditionally the third mode of perseity has been taken to be the manner in which a primary substance exists in itself and not in another. This is not significant for our theory of demonstration.

true. From another point of view, demonstration is either a priori or a posteriori. A priori demonstration moves from a necessary cause to its effect, and a posteriori demonstration proceeds from an effect to its necessary cause. Combining the dividing members of these two divisions of demonstration, it would seem perhaps that there are four types of demonstration, namely, the a priori explanatory, the a posteriori explanatory, the a priori demonstration of fact, and the a posteriori demonstration of fact. However only three of these four are real possibilities. An a posteriori explanatory demonstration is not possible. The existence of an effect can point to the existence of a cause but it can never explain the cause, for the cause is, by definition, the explanation of the effect. Let us consider the three types of demonstration one by one, beginning with the most perfect type, that is, the a priori explanatory demonstration. Since there can be no a posteriori explanatory demonstration it is not necessary in naming the most perfect type to say any more than "explanatory demonstration."

A. *Explanatory Demonstration*

The name "explanatory demonstration" is reserved for the most perfect type of demonstration in which a conclusion is proved through a middle term which adequately expresses that on account of which (*propter quid*) the conclusion is what it is and must be as it is. Formally, an explanatory demonstration is a first figure categorical syllogism concluding directly. Materially, the middle term of the explanatory demonstration is related to the major term as a necessary and proximate cause to a convertible effect. There are three different ways to construct an explanatory demonstration:

1. The first is to prove an immediate property of a subject through a middle term which is simultaneously the definition of the subject and the proximate cause of this property. This type of demonstration can proceed by way of a middle term which defines in terms of any one of the four causes (material, formal, efficient, and final) for natural science, of one cause (formal) for mathematics, and of any one of three causes (formal, efficient, and final) for metaphysical science, e.g.:
   Every rational animal is capable of speech.
   But every man is a rational animal.
   Therefore, every man is capable of speech.

We have already analyzed this example (pp. 259, 264). Summarizing our findings briefly, we can make the following observations. Formally this is a first figure categorical syllogism concluding directly. It is a valid syllogism with an AAA *mood*. On the side of matter: The minor term is a scientifically knowable subject. The major term is a specific property of the scientific subject. The middle term is the essential definition of the subject and the proximate and necessary cause of the property. Each proposition in the demonstration is universally necessary, and each is convertible. The major premise is a self-evident proposition in the fourth mode of perseity. The minor premise is a self-evident proposition in the first mode of perseity. The conclusion is a mediate proposition (shown to be necessarily true through the premises) in the second mode of perseity.

2. The second is to prove a mediate property[2] of a subject through a middle term which is an immediate property of the subject and the proximate cause of the mediate property, e.g.:

> **Everything having intellect and imagination can learn geometry.**
> **But every man has intellect and imagination.**
> **Therefore, every man can learn geometry.**

3. The third is to prove one cause of its subject through a middle term expressing another cause of this same subject when the former cause is related to the latter as effect to cause, e.g.:

> **Every argumentation productive of science is from premises which are ultimately necessary and immediately true.**
> **But every demonstration is an argumentation productive of science.**
> **Therefore, every demonstration is from premises which are ultimately necessary and immediately true.**

In this example, a definition of demonstration itself in terms of material cause is demonstrated by way of a middle term which defines demonstration in terms of final cause. The causality of the material cause depends on the prior three causes. Hence, the material cause can be demonstrated by the final, efficient, or formal cause. The causality of the formal cause depends on the efficient and final causes. Hence,

---

[2] Some properties find their proximate cause in the principles intrinsic to the essence of their subjects. A second type of property finds its proximate cause in a property of the first type and belongs to the essence through this prior property. The first type of property is immediate; the second type is mediate. The example of a mediate and an immediate property used here is borrowed from the translators of *The Material Logic of John of St. Thomas*, translated by Yves R. Simon, John J. Glanville, and G. Donald Hollenhorst (Chicago: University of Chicago Press, 1955), p. 617. The footnotes supplied by the translators to Chapter 6, pp. 616–625, are recommended as extremely illuminating for the theory of demonstration.

the formal cause can be demonstrated by either efficient or final cause. And the efficient cause depends upon the final cause, so that the efficient cause is able to be demonstrated by way of the final cause.

B. A Priori Demonstration of the Fact

An a priori demonstration of the fact proves that the predicate of the conclusion belongs to its subject by relating them through a middle term which expresses the remote or mediate cause of the predicate. In so far as the middle term is not the proximate cause of the effect, this falls short of the perfection demanded for explanatory demonstration. But in so far as this is a demonstration from cause to effect, this type of demonstration is a priori. There are two ways to construct an a priori demonstration of fact. In the first case the cause through which the effect is proved is remote in the sense of being once removed in the real order. In the second case the effect is proved through a cause remote in the sense of being generic, that is, once removed in the order of abstraction.

1. An effect is demonstrated by way of a remote or mediate cause whenever a mediate property is proved of its subject through a middle term which defines this subject. Since the definition of the subject expresses the proximate cause of the immediate property in question, this definition expresses only the remote or mediate cause of the property proved to belong to the subject in this demonstration, e.g.:

   > Every rational animal can learn geometry.
   > But every man is a rational animal.
   > Therefore, every man can learn geometry.

2. A significantly different way to prove an effect by way of a remote or mediate cause is in a demonstration which proves an effect from a cause which is mediate and nonconvertible, but exclusive (exclusive in the sense that, though it is not universally a cause of the effect, the effect cannot be achieved save that it function as a generic cause), e.g.:

   > Everything capable of speech is an animal.
   > But no plant is an animal.
   > Therefore, no plant is capable of speech.

The first kind of a priori demonstration of fact is expressed in the first figure concluding directly. However, the second type can be expressed only in the second figure, and, accordingly, allows only for a negative conclusion. The reason for this is that the effect proved through the remote cause is convertible with that cause in the first type of a priori demonstration of fact, while it is less extended than the cause in the second type. Any attempt to argue affirmatively in the second instance results in an unextended middle term.

C. A Posteriori Demonstration of Fact

An a posteriori demonstration of fact is achieved by way of a middle term which is related to the predicate of the conclusion as effect to cause. One instance of this type of demonstration involves an effect convertible with its cause, while another involves an effect less extended than its cause.

1. The first instance of an a posteriori demonstration of fact is a syllogism proving a cause through an effect convertible with this cause. Examples of this type of demonstration can be achieved by interchanging the middle and major terms in examples of explanatory syllogisms, e.g.:

   > **Whatever is capable of speech is a rational animal.**
   > **But every man is capable of speech.**
   > **Therefore, every man is a rational animal.**

   We have said that demonstration is an analogous notion, which is perfectly realized only in explanatory demonstration. The requirements for the less strict types of demonstration are not identically what they are for explanatory demonstration. In this example of an a posteriori demonstration of the fact we have a fine illustration of this point. This example is like the first example offered for the explanatory demonstration, except that here the middle and major terms of that example are interchanged. In our first example foreknowledge of the essential definition of the scientific subject is a necessary prerequisite for demonstration. However, in this present example only the nominal definition of the subject is presupposed. To know the essential definition of the subject here is already to know the conclusion, which simply predicates this definition of the subject.

2. A second kind of a posteriori demonstration of fact proves a cause through an effect which is less extended than, but yet exclusive to, that cause, e.g.:

   > **Whatever is capable of speech has a soul.**
   > **But every man is capable of speech.**
   > **Therefore, every man has a soul.**

   We should point out that any attempt will be logically defective which seeks to prove a cause of a subject through an effect less extended than that subject or through an effect which is not exclusive to the cause. In the first case the syllogism attempted has an illicit minor term, and in the second case it has an unextended middle term.

Each of the types of demonstration treated admits of a form weaker than the one described and illustrated above. For each there is a weaker form in which the middle term remains the same while the minor term becomes a subjective part of the original minor. This is a weaker form since neither the minor premise nor the conclusion is commensurately universal. In addition, the explanatory demonstration and the a priori demonstration of the fact which proceeds from the definition of the scientific subject to a mediate property of this subject admit of a weaker form which employs the original minor term as middle and a subjective part of the original minor term as minor. Consider, for example, the following explanatory demonstration (No. 1) and its weaker forms (Nos. 2 and 3):

1. Every argumentation productive of science is from premises which are ultimately necessary and ultimately true.
   But every demonstration is an argumentation productive of science.
   Therefore, every demonstration is from premises which are ultimately necessary and ultimately true.
   (The conclusion is convertible, and the middle is proximate both to the minor term and the major term.)

2. Every argumentation productive of science is from premises which are ultimately necessary and ultimately true.
   But every explanatory demonstration is an argumentation productive of science.
   Therefore, every explanatory demonstration is from premises which are ultimately necessary and ultimately true.
   (The conclusion is not convertible, and the middle is not proximate to the minor term although it is proximate to the major.)

3. Every demonstration is from premises which are ultimately necessary and ultimately true.
   But every explanatory demonstration is a demonstration.
   Therefore, every explanatory demonstration is from premises which are ultimately necessary and ultimately true.
   (The conclusion is not convertible, and the middle is not proximate to the major term although it is proximate to the minor.)

VII. The Middle Term in Demonstration

The middle term in the scientific syllogism is the most significant single principle of demonstrative discourse. The fundamental task of anyone who attempts to demonstrate is to find a middle term which can manifest the necessary connection between the predicate and the subject of a scientific conclusion. This middle term must in every case be the cause of the knowledge of the connection between the extremes in the conclusion. In an a posteriori proof the middle term is a cause only of the knowledge of the connection. In a priori proof it is both a cause of the knowledge of the connection and the real cause of that connection. In an explanatory demonstration the middle term is the adequate and proximate real cause of the connection between the predicate and subject of the conclusion. In an a priori demonstration of fact the middle term is a real cause, but a cause which is either less than adequate (because only generic) or less than proximate (because once removed) in reference to the effect in question.

A more penetrating study of demonstration than that suited to an introductory logic course would demand a detailed consideration of middle terms and the modes of defining proper to them. We shall, however, only sketch a plan of inquiry suited to a more advanced study. The most significant instance of a middle term is the definition of the scientific subject employed in the explanatory demonstration. The force of definitions in scientific discourse differs to the degree in which the definitions differ in the way they abstract from matter. From this point of view, there are three generically distinct modes of defining, and each determines a generically distinct scientific discipline. The three genera of science are respectively natural science, mathematical science, and metaphysical science. Demonstrations in the natural sciences have instances of physical or sensible (i.e., changeable) being as a subject of inquiry, and they involve middle terms which define their subjects in the only way these subjects can exist outside the mind, namely, in matter. *Man*, for example, is defined in terms of "flesh and blood," and no man exists outside the mind save as concretized in flesh and blood. The subjects of mathematical demonstrations are ultimately instances of quantified being, and the middle terms in mathematical demonstrations define their subjects without matter, even though they can exist outside the mind only in matter. Thus, *triangle* is defined in abstraction from any physical characteristics, although triangles can exist outside the mind only as concretized in triangularly shaped physical things. The subject of metaphysical demonstration is ultimately being itself, and the middle term in metaphysical demonstration defines its subject in a way in which this subject can exist outside the mind, that is, without matter. Being is in some cases material, but being need not be material, and, in fact, is not material in God. We can distinguish between the subjects proper to physical science, mathematics, and metaphysics by noting that the first is concerned with beings dependent on matter both for their existence and for their intelligibility; the second, with beings dependent on matter for their existence but not for their intelligibility; the third, with beings independent of matter both for their existence and their intelligibility.

It is not the business of logic further to pursue the differences between the various sciences, nor to discuss the *peculiar* characteristics of demonstration in the different sciences. It is the business of logic to investigate the general notion of demonstration and generally its division into types. Each particular science has its own proper methodology. To the extent to which a particular science demonstrates conclusions, it contracts the general notion of demonstration to the needs of its own proper subject matter. The general theory of demonstration is open to contraction in the face of the scientific subjects respectively proper to the several sciences. This is true whether the subject be proper to the natural sciences, mathematics, or metaphysics. The general theory must not be confused with any one of its contractions.

VIII. Dialectical Argument

The strictest form of argument is the demonstration, which yields a certainly true conclusion proven in the light of necessarily true premises. However, a demonstration is comparatively rare. For one thing, demonstration bears only upon necessary matter, and many items even of academic significance are contingent rather than necessary. This is particularly true, for example, in the area of political action. For another thing, the human intellect is frequently limited in its attempts to define a thing and consequently limited in reference to what it might demonstrate of that thing. For example, demonstrations from essential definition to property are impossible for natural species save the species *man*, since man is the only natural species we can essentially define. Perhaps the most favorable area for examples of demonstrative argumentation is that of classical mathematics. Mathematical entities are not subject to the contingent attributes which are characteristic of natural things, and they are relatively easy for the human intellect to understand. Nevertheless, even in mathematics demonstration does not come without some difficulty.

There is a second form of argumentation which is not demonstrative, but which is nevertheless legitimate, and, in fact, significantly so. This kind of argumentation, called dialectical, falls short of a certainly true conclusion, but it does give conclusions at least probably true. A proposition is certainly true when the evidence for it rules out the possibility of its contradiction. A proposition is probably true when there is solid evidence for it, but when this evidence leaves open the possibility of its contradiction. A certainly true conclusion is more excellent than a probably true conclusion. However, a probably true conclusion is better than no conclusion. Probable argumentation has a significant place both as a substitute for, and as a supplement to, demonstration.

Generally speaking, the name "dialectic" can be applied to any form of logical discourse which is less than demonstrative. This includes the kind of discourse which is characteristic of rhetoric and poetry. In a more restricted sense dialectical discourse is divided off against demonstration on the one hand and rhetoric and poetry on the other. From this point of view the following points can be made. Demonstration aims at science understood as certain knowledge proven through causes. This involves a commitment to one side of a contradiction without fear of the opposite. Dialectical argumentation aims at opinion. This involves a commitment of the mind to one side of a contradiction, but with a fear of its opposite. The rhetorical argument intends to persuade one to accept one side of a contradiction. Poetry, in so far as it can be considered an instrument for teaching, seeks to generate an inclination to one side of a contradiction through a pleasing representation of it. We shall continue to speak of dialectics without reference to rhetorical argumentation or poetry. The scientist who demonstrates his conclusion resolves one of two contradictory propositions into self-evident principles which demand acceptance of this proposition and the rejection of its contradictory. The dialectician, on the other hand, resolves one side of a contradiction into premises which certainly support it as probable but which allow for other premises which might support its opposite. On the side of form, the demonstration and the dialectical syllogism are of equal status — each must be, for its own purposes, indefectibly valid. It is on the side of matter that they differ. The demonstration proceeds from premises which cannot be false, because they are ultimately self-evident. The dialectical argumentation proceeds from probably true premises, which can, of course, communicate to their conclusions no more than their original probability. Some of the more usual dialectical premises include those in contingent matter which are true for the most part but which admit of exceptions, those which are probably true either because they are commonly held by most men or are proposed by some respected authority on the point, and those which are proper to one field but which lend themselves as argumentative principles in another. Each of the following argumentations is at best dialectical:

1. Mothers love their children. (Probable in the sense of being true for the most part.)
   But Susan is a mother.
   Therefore, Susan loves her children.

2. All contemporary natural species have evolved from less perfect species. (Probable because it is the opinion of experts in the field of biology.)
   But man is a contemporary natural species.
   Therefore, man has evolved from a less perfect species.

3. Contraries are within the same genus. (Certain in logic and applicable in psychology to yield a probable conclusion.)
   But love and hatred are contrary passions.
   Therefore, love and hatred are passions of the same appetite.

Dialectic differs from a demonstration in that the former is unterminated while the latter is terminated. Demonstration is terminated because its conclusion is resolved into propositions which infallibly represent reality. Dialectical conclusions are not certain because they are not resolved into propositions which infallibly represent real things. Dialectic remains within the mind while tending to the real. For this reason the dialectician must be able to employ every logical device calculated to manifest the truth of a proposition. As a dialectician he cannot see the necessary correspondence between his conclusions and reality. Thus he must be prepared to explore every avenue of logic open to him in order to come as close as possible to a certainly true conclusion, although he can never reach certitude through dialectic. He must be skilled in the arts of defining and dividing. He must be able to distinguish between terms on the basis of the most subtle differences in signification and supposition. He must be able to classify predicates predicably and categorically so that the meanings of terms and the force of propositions may be fully evaluated. He must be able to distinguish between propositions on the basis of formal differences, and, even more so, on the basis of material differences. He must know the various forms of argumentation and be aware of the properties determining their probative force. He must know the minimum limits for scientific discourse and be especially skilled in arguments falling between this limit on the one hand and illogical discourse on the other. He can in a sense be less rigid and more flexible, more general and less pointed, less rigorous perhaps, but nonetheless never less logical, than the one who demonstrates. He is satisfied with less, for certainty is more satisfying than probability. But he is more often satisfied, since many more things can be shown dialectically to be probable than can be shown demonstratively to be certain. The course in logic has been ordered primarily to demonstration because the scientific syllogism is the prime instance of logical discourse. It has also been ordered to dialectical argumentation, in the realization that the academic intellect does not live on demonstration alone but looks to dialectics as a necessary complement in the quest for truth.

Perhaps the most significant use of the dialectical method in our day is its application as "scientific method" in the empiriological disciplines. We have talked about the several uses of the name "science." In the language of most people the term is applied to those disciplines such as physics, chemistry, and biology, which can be called empiriological. Strangely enough, the method which characterizes these disciplines is nonscientific in the Aristotelian sense. It is, as we have used the term, dialectical rather than scientific. We should not think that there is one method univocally applied to the subjects respectively proper to the several empiriological disciplines. Yet there is generally one method applied most strictly in physics and proportionally in the other empiriological disciplines which can be spoken of as the "scientific method." In its most simple outline the method involves first of all the formation of a hypothesis or theory calculated to explain what has been observed in a given physical, chemical, or biological area of interest. The method allows as truly significant only those data which can in some way be quantitatively measured, and it involves a hypothesis which is always mathematical in form even though it bears upon the physical. From the hypothesis — which as originally formulated explains only what has already been observed — conclusions are deduced which are translated into predictions of results from future experiments. When the results which are predicted are in fact realized, the hypothesis is said to be verified. It is too much here to claim that the hypothesis has been established as certainly true. In effect, it has been shown only to be a possibility. Its verification points to probability at best. If it is subsequently verified in a different experiment, it is shown to be more probable. However, it can never be shown to be more than highly probable no matter how many times and in how many ways it is experimentally verified. This is a good example of the unterminated dialectical procedure which generates a probable conclusion. Let us look more carefully at the "scientific method" in order to see that this is the case. Given the hypothesis as an antecedent, the predicted results are related to it as a conclusion validly deduced from its premises. In the light of the experiment this conclusion is seen certainly to be true (after the fashion of a factually evident proposition). Now, as we have seen, a true conclusion can come either from true or false premises. Thus, beginning with the conclusion as evidence, all we can say of the premises is that they could be true. In the demonstrative argumentation the premises are better known than the conclusion. Given as certainly true, they generate a certainly true conclusion. In the empiriological argumentation the conclusion is better known than the premises. Given as true, it can do no more than indicate the possibility of the premises. Thus, the method of empiriological science is at best dialectical. It can never establish any of its hypotheses as necessarily true. Although it lacks the perfection of the demonstrative method as far as certainty is concerned, it is much more widely applicable than the demonstrative method. It has yielded and it will continue to yield significant, though only probable, knowledge about a multitude of things which are more or less impervious to the method of demonstration. The logical character of the empiriological method is a clear illustration of dialectical argumentation. Its widespread and profitable employment is an example of the intrinsic worth of the dialectical procedure.

We have already noted that the demonstrative method presupposes prior logical discourse which paves the way for the demonstrative proof. In the sense that this predemonstrative discourse is nondemonstrative it can be spoken of as dialectical. In this sense of the term, dialectical discourse can prepare for demonstration in two significantly different ways. It does this in a negative fashion when it is employed to refute false opposing positions. It does this positively by constructively paving the way for demonstrative argumentation. It may be that much more effort must be expended dialectically in preparation for a demonstration than in the demonstration itself. We have mentioned this point already, to forestall impatience on the part of the student in the face of the apparent anticlimactic character of demonstration. Once the essential definition of the scientific subject in a strict explanatory demonstration is known, once the nominal definition of its property is known, and once the self-evident character of the premises of the demonstration is appreciated, the conclusion is easily and quickly seen as demonstrably guaranteed. Yet there may be long and difficult processes of division, definition, and argumentation necessary before the principles necessary for the demonstration in question are achieved. It is not that an essential definition cannot logically stand on its own two feet, nor that a self-evident proposition depends logically on prior propositions. This would be absurd. Yet a definition is not easily understood just because it is essential, nor is a proposition easily grasped because it is self-evident. There usually is the need for serious and complicated discourse in order to understand fully a definition or a self-evident proposition. Once the definition or the self-evident proposition is grasped, the dialectical scaffolding erected so that it might be grasped can be torn down, and the definition or the self-evident proposition will stand alone. Yet, without the dialectical scaffolding, perhaps neither would have ever been grasped.

Thus dialectical discourse serves at least three purposes. It is of predemonstrative value in so far as it helps to pave the way for demonstration where demonstration is possible. It complements the demonstrative method by supplying arguments of subsidiary value in areas where some demonstration is possible. And it serves in place of demonstration in many areas where demonstrative procedure is out of the question for one reason or another.

Exercise XIX

1. Define: science (empiriological), science (in Aristotle's sense), demonstration, material universality, perseity, commensurate universality (or convertibility), explanatory demonstration, demonstration of the fact, a priori demonstration, a posteriori demonstration, dialectical argumentation, rhetorical argumentation, poetical argumentation, "scientific method."
2. What is the difference between the formal and the material consideration of the syllogism? Illustrate.
3. What is science in Aristotle's sense? What is the connection between this and demonstration?
4. What does it mean to say demonstration is analogous?
5. What must be known prescientifically before one is able to construct a proof which is an instance of the strictest type of demonstration?
6. Explain how it is that the definition of demonstration in terms of material cause is necessitated in the light of its definition in terms of final cause.
7. List, and explain, the requirements for the premises of a strict demonstration.
8. What is the role of the self-evident proposition in the theory of demonstration?
9. *Every man is a rational animal* is materially as well as formally universal; it is in the first mode of perseity; and it is commensurately universal or convertible. Explain. What about *Every man is animal?* What about *Every stone is colored?*
10. Demonstration is logically divided into types by way of two distinct dichotomies. Nevertheless, there are only three basically different kinds of demonstration. Explain.

11. What are the various ways in which explanatory demonstration can be achieved? Explain what it is about each which keeps it on the level of explanatory demonstration.

12. In what way is an a priori demonstration of fact more like an explanatory demonstration than an a posteriori demonstration? In what way is it less than explanatory? How do the two remote causes, respectively distinguishing the two ways of demonstrating in an a priori demonstration of fact, differ significantly from one another? What do they have in common so that each is a principle of basically the same type of demonstration?

13. When is a nonconvertible effect able to be a middle term in demonstration? When is a nonconvertible effect an illicit principle in demonstration? Illustrate the difference between an a posteriori demonstration of fact by way of convertible effect on the one hand and a legitimate nonconvertible effect on the other.

14. Why is it accurate to say that the middle term in the scientific syllogism is the most significant single principle of demonstrative discourse?

15. In general how are differences in sciences determined according to different modes of defining?

16. What is the relation of the logic of demonstration to the methodologies respectively of the particular sciences?

17. Distinguish between demonstration and dialectical discourse. Carefully explain the relationships between them.

18. What are the advantages and disadvantages of demonstrative argumentation on the one hand and dialectical argumentation on the other?

19. Suggest and explain several examples of dialectical discourse.

20. Discuss the "scientific method" and the manner in which it differs from the method of demonstration. Explain the reasons why it is a good example of dialectics.

21. The following argumentations are suggested as illustrations in a well-known treatise on demonstrative discourse entitled *On Demonstration*.[3] Examine each one carefully to see if you can discover the logical points each is supposed to illustrate.
    21.1. Every plane figure bounded by three lines, having an exterior angle equal to two opposite interior angles, has three angles equal to two right angles.
    But every triangle is a plane figure bounded by three lines, having an exterior angle equal to two opposite interior angles.
    Therefore, every triangle has three angles equal to two right angles.

---

[3] A translation of this treatise by Vincent R. Larkin is found in *Readings in Logic*, edited by Roland Houde (Dubuque: William C. Brown Company, Publishers, 1958), pp. 148–152. The *De Demonstratione* is a brief work on the theory of demonstration, traditionally listed among the works of St. Thomas Aquinas. Although it is uncertain that St. Thomas is the author, it is agreed that the teaching is authentically Thomistic.

21.2. Every natural body illumined by the sun when deprived of light by the interference of the earth is eclipsed.
But the moon is a natural body illumined by the sun and sometimes deprived of light by the interference of the earth.
Therefore, the moon is sometimes eclipsed.

21.3. Every triangle has three angles equal to two right angles.
But every isosceles figure is a triangle.
Therefore, every isosceles figure has three angles equal to two right angles.

21.4. Every plane figure bounded by three lines, having an exterior angle equal to two opposite interior angles, has three angles equal to two right angles.
But every isosceles figure is a plane figure bounded by three lines, having an exterior angle equal to two opposite interior angles.
Therefore, every isosceles figure has three angles equal to two right angles.

21.5. Every luminous nontwinkling body is situated near the observer.
But a planet is a luminous nontwinkling body.
Therefore, a planet is situated near the observer.

21.6. Every luminous body which is situated near the observer is nontwinkling.
But a planet is a luminous body which is situated near the observer.
Therefore, a planet is nontwinkling.

22. The following examples are taken from a scholarly article on demonstration written by William A. Wallace, O.P.[4] Examine each one carefully; then classify and evaluate it in the light of what we have seen about demonstration.

22.1. What is done always or for the most part, and for what is best and most suitable, is for a determined goal or for an end (i.e., cannot be merely by chance).
But what is done by nature is done always or for the most part, and for what is best and most suitable.
Therefore, what is done by nature is for a determined goal or for an end.

22.2. A thing divisible into parts on which depends the motion of the whole is moved by another.
But whatever is in motion is a thing divisible into parts on which depends the motion of the whole.
Therefore, whatever is in motion is moved by another.

22.3. A moving thing that is moved by at best a finite series of moved movers is moved by a first Mover completely unmoved.
But whatever is moved is a moving thing that is moved by at best a finite series of moved movers.
Therefore, whatever is moved is moved by a first Mover completely unmoved.

---

[4] "Some Demonstrations in the Science of Nature," *The Thomist Reader*, 1957, pp. 90–118.

22.4. A body that waxes and wanes through crescent phases is a spherical physical body illumined by an external source (i.e., the sun).
But the moon is a body that waxes and wanes through crescent phases.
Therefore, the moon is a spherical physical body illumined by an external source (i.e., the sun).

22.5. An organism that (not having a photosynthetic process to make its own food) must seek food from its surroundings is an organism endowed with sensation.
But an animal is an organism that must seek food from its surroundings.
Therefore, an animal is endowed with sensation.

22.6. A fluid of limited quantity that flows continuously in one direction is moved circularly.
But blood is a fluid of limited quantity that flows continuously in one direction.
Therefore, blood is moved circularly.

22.7. A rational animal is capable of natural science.
But man is a rational animal.
Therefore, man is capable of natural science.

22.8. A being who performs immaterial cognitive acts is a being with an immaterial cognitive faculty.
But man is a being who performs immaterial cognitive acts.
Therefore, man is a being with an immaterial cognitive faculty.

22.9. A being with an immaterial cognitive faculty is a being with an immaterial substance that is the subject of that faculty.
But man is a being with an immaterial cognitive faculty.
Therefore, man is a being with an immaterial substance that is the subject of that faculty.

22.10. A spiritual substance or an immaterial form is incorruptible or immortal.
But the soul of man is a spiritual substance or immaterial form.
Therefore, the soul of man is incorruptible or immortal.

23. Examine argumentations found in your textbooks and reference books. Evaluate them from both the formal and the material point of view, especially distinguishing the strict proofs from the dialectical argumentations.

CHAPTER XX

Induction

I. The Insufficiency of Deduction

The basic instance of deduction is the categorical syllogism which shows how the extremes in the conclusion are related through a middle term which appears in each of the premises. The connection between this middle term and the other term in each premise may be evident apart from a further middle term or it may itself be manifested by way of another middle term. However, it is impossible to admit an infinite series of middle terms. Sooner or later the deductive conclusion of the categorical syllogism must be resolved into premises which are evident in the light of something other than deductive evidence. We have seen in an earlier chapter that at least one premise in a categorical syllogism must be universal, and truly universal, not merely enumeratively so. What is the character of this universal proposition which is necessary as a principle of syllogistic argumentation but cannot be its term? Where does it come from, and what is the guarantee of its adequacy?

To be more specific, let us consider the most significant instance of the categorical syllogism, namely, demonstration. The conclusion in a demonstration is proved to be necessarily true in the light of premises which are necessarily true and manifest the truth of the conclusion. These premises may be seen to be necessarily true because they are proved to be true in the light of previous premises. However, unless some propositions are necessarily true without proof, then no premises at all can prove any conclusion. An endless series of proved premises is absurd. Demonstrated conclusions are resolved into premises which are ultimately principles of proof without being proved themselves — or there are no demonstrations at all. What is the character of these nonproved principles of proof? Where do they come from, and what is the guarantee of their adequacy?

The answer to these questions leads us to a discussion of induction.

The universal propositions which serve ultimately as principles of deductive argumentation are seen in the light of their singular instances. The human intellect is able, so long as there is sufficient sensory experience, to rise from the singular instances of a universal proposition to an understanding of that proposition itself, either because of a sufficient enumeration of singular instances or because of an insight into the meanings of the terms involved. In either case the ascent of the mind from the singulars to the universal is spoken of as induction. In general the universal principles of deduction are products of an inductive movement of the mind from singular instances of a truth to a universal expression of this truth which is guaranteed as adequate because it is grounded in an undeniable experience of singulars. In the special case of demonstration, as we shall see, the nonproved principle of proof is a self-evident proposition, which is in its own way the product of an induction.

II. The Types of Induction

We have said that the movement of the mind from singular instances of a universal proposition to that universal proposition is induction. As a matter of fact, the notion of induction can be extended to mean any passage of the mind from the particular or singular to the universal. This would include the abstraction of the universal object of simple apprehension from the singulars in which it is concretized as well as the induction of the universal proposition from its singular instances. Induction, generally understood, terminates either in the incomplex object of the first operation or the complex object of the second. We have said, and we shall soon see why this is so, that induction terminates in the complex object of judgment either as the result of an insight into the meanings of its terms or as a result of a sufficient enumeration of singular instances of the universal. The first of these last two inductions is immediate, while the second is an example of reasoning.

Thus we can distinguish between the following three types of induction:

1. **Abstractive induction** — the movement of the mind from the concrete instances of a nature to an understanding of that nature taken absolutely — achieved on the level of simple apprehension.
2. **Immediate induction of self-evident propositions** — the movement of the mind from an understanding of the terms involved in a universal proposition to the necessary truth of that proposition itself — achieved on the level of the second operation of the intellect in an immediate judgment.

3. **Mediate induction** — the discursive movement of the mind from a sufficient enumeration of the singular instances of a universal proposition to the truth of that proposition — achieved in the conclusion to an inductive reasoning process.

The subject of abstractive induction has been treated elsewhere in this book. There is no need to go into it again at this point. However, it may be worthwhile for the student to review what has already been said on the subject, especially in Chapter II and in Chapter IV. The second kind of induction, namely, the immediate induction of the self-evident proposition, is so proximately involved with abstractive induction that it cannot be understood unless abstractive induction itself is clearly grasped. For this reason, serious reconsideration of what has already been said on abstractive induction is recommended as a preliminary to the following sections on the induction of the self-evident proposition.

III. The Self-Evident Proposition

We have seen that the premises in a scientific syllogism must be ultimately indemonstrable principles of demonstration. That is, they must be seen to be universally and necessarily true in themselves and apart from any propositions brought to bear upon them as evidence for their truth. They must be premises in an absolute sense, namely, principles of proof which are themselves unproved precisely because they cannot and need not be proved. Propositions of this type must be self-evident propositions, that is, propositions which are known certainly to be necessarily true once their terms are understood.

The subject and predicate of a self-evident proposition exist in necessary matter and are so proximately connected with one another that an understanding of them compels one to assent to the necessary truth of the proposition in which they are joined, e.g., the proposition *The whole is greater than any one of its parts*. No one can understand the meaning of the terms in this proposition without immediately seeing the absolutely necessary truth of the proposition itself.

The self-evident proposition is an immediate proposition. This means that it does not require prior propositions to make it evident. However, not all immediate propositions are self-evident. We have noted before that a proposition which finds its evidence in an undeniable factual situation is also immediate. We have spoken of this second type of immediate proposition as factually evident in order to distinguish it from the self-evident proposition. The factually evident proposition is a report simply on what happens to be the factual situation. As such, the truth of a factually evident proposition is guaranteed by a factual situation, which, as factual, might possibly change, e.g., *The weather is pleasant today*. Assent to a self-evident proposition is an irrevocable assent to a necessary connection between a subject and its predicate, in the face of the absolute impossibility of an assent to its contradiction. Assent to a factually evident proposition is demanded by what happens to be the case, and by what perhaps need not and possibly later on will not be the case. Neither the self-evident nor the factually evident proposition depends upon prior propositions to manifest the connection between its extremes. The very meanings of the terms involved does this for the self-evident proposition, and the factual situation immediately present to the senses does this for the factually evident propositions. Both are immediate propositions. However, only the self-evident proposition can be understood to be an adequate absolute premise in demonstration. The factually evident proposition may find its way into a less strict instance of demonstration. But one premise, at least, in every demonstration must be ultimately self-evident. The necessity called for in a scientific proposition can be communicated to a conclusion only by a proposition which is immediate and irrevocably necessary. This much is true of the self-evident proposition and only of the self-evident proposition.

We began our discussion of the self-evident proposition by noting the need for an indemonstrable proposition as an absolute starting point for demonstration. The self-evident proposition satisfies this need. It is, first of all undemonstrable. However, not every undemonstrable proposition is self-evident. All propositions in contingent matter are undemonstrable, of course, since demonstration bears only upon the necessary. However, propositions in contingent matter are undemonstrable because of a deficiency in matter. Self-evident propositions are always in necesary matter, and their indemonstrability springs rather from their excellence than from any inherent deficiency. Demonstration makes evident what is not already evident. If something is demonstrable it suffers privation, since it lacks evidence and must find such evidence in premises. Self-evident propositions do not suffer this privation, since they are already evident in themselves.

Self-evident propositions are aptly described as the basic truths of the scientific syllogism. They are *basic* truths in so far as they are universally and necessarily true while admitting of no prior propositions necessary to make them evident. They are truths *of the scientific syllogism* in so far as they are principles from which conclusions which have the character of scientific propositions are generated.

IV. The Types of Self-Evident Propositions

Self-evident propositions fall into different classes. Some are said to be self-evident only in themselves. Others are said to be self-evident to us. Of these latter some are said to be self-evident in an unrestricted sense. These are self-evident to all of us. Others are said to be self-evident only to those skilled in a given scientific area. These are self-evident to the learned. Let us consider the principles according to which the self-evident proposition is divided and subdivided in this fashion.

A self-evident proposition is a proposition known certainly to be true once its terms are understood. The most perfect instance of this is found in the proposition in which the predicate is the essential definition or an element in the essential definition of the subject.[1] Once the subject is defined, the identity of subject and predicate is grasped and the intellect is moved to commit itself irrevocably to the truth of the proposition. Suppose, however, that the predicate of a proposition lies within the definition of its subject, but suppose that this subject is so difficult to know that it cannot be defined essentially. Such a proposition would be self-evident *in itself*, but it would not be self-evident to us. If the predicate of a proposition lies within the definition of its subject and if its subject can be essentially defined, this proposition is not only self-evident in itself but also self-evident *to us*. If the subject of this proposition were a common notion, that is, one which can be understood by all men, the proposition would be self-evident *to all of us*. If the subject could be defined only by those specially equipped intellectually to operate within a limited field, the proposition would be self-evident only to those skilled in that field, that is, self-evident only *to the learned*. Thus, it is not too difficult to see, at least in reference to the primary instance of the self-evident proposition, the reasons behind the division and subdivision of the self-evident proposition with which we began this section.

In defending his argument for the existence of God St. Thomas appeals to the fact that the proposition *God is* is not self-evident to us, even though it is self-evident in itself. Were we to know the essence of God,

---

[1] This instance of the self-evident proposition can be seen to be identified with a proposition in the first mode of perseity. However, it is not to be understood that the notion of self-evident proposition (even though these are traditionally referred to as *per se nota* propositions) is to be identified with the notion of a proposition involving a mode of perseity (traditionally called a *modum dicendi per se*). The conclusion in a demonstration proving a property of its proper subject is in the second mode of perseity without, of course, being self-evident, since it is a conclusion.

we could not — nor would we need to — demonstrate His existence, for His essence is His existence. Yet, since we do not fully comprehend His essence, this proposition is not self-evident *to us*. We must, and can, make it evident by proving that God exists by means of an a posteriori demonstration which argues from His effects, which are known to us, to their cause — God Himself. Aristotle and St. Thomas both supply us with examples of propositions which are self-evident to all of us. These propositions are known to all because their terms are common notions easily and surely grasped by all men. The examples include: *The same thing cannot be and not be. The same proposition does not admit simultaneously of affirmation and denial. The whole is greater than any one of its parts. Things equal to one and the same thing are equal to one another.* Propositions of this type can be spoken of as axioms. Axioms must be known and assented to by anyone who engages in rational discourse. They are the absolutely ultimate and common principles which guarantee the integrity of all discourse and into which all discourse is in one way or another resolved. Propositions which are self-evident only to the learned are related to the axioms as the proper to the common. They are known only to a limited number of men precisely because the terms involved are less common and more determinate. St. Thomas illustrates the more restricted type of self-evident proposition by suggesting the proposition *All right angles are equal.* This is a proposition immediately evident only to one who knows that equality enters into the definition of *right angle*, and the definition of *right angle* is a definition which escapes the knowledge of many. Another example traditionally offered is the proposition *Incorporeal substances are not situated in place.* We can add to these any proposition in which the essential definition or part of the essential definition is predicated of a specific subject, such as *Every man is a rational animal.* A proposition of this type is sometimes spoken of as a thesis rather than an axiom. We saw that the axioms are necessary if we are to demonstrate in any scientific area. A thesis proper to a given area is necessary for demonstration properly within that area. Axioms may or may not be used explicitly as premises in a demonstration. Nevertheless, the principle of contradiction, namely, *The same proposition does not admit simultaneously of affirmation and denial,* is presupposed necessarily as a methodological principle of any demonstration whatsoever (i.e., as a principle guaranteeing the logical integrity of the demonstration) even though it is not used explicitly as a premise. For no thesis can function properly as a principle of demonstration unless it is firmly understood that its denial and the affirmation of its opposite are excluded in the face of its own affirmation.

Before leaving this section let us recapitulate by considering the following schema which represents the division and subdivision of self-evident proposition into its types.

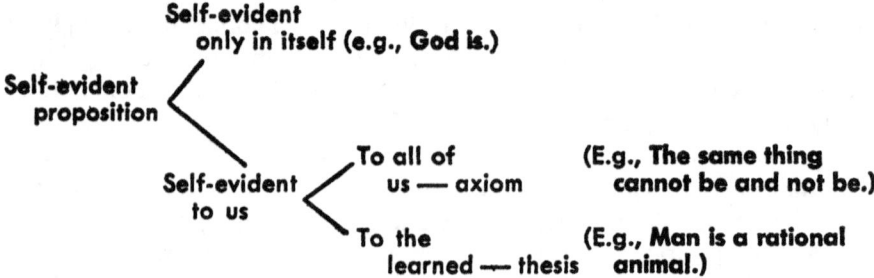

V. The Induction of Self-Evident Propositions

Self-evident propositions are, in a sense, natural to the human intellect. This does not mean that they are innately possessed and fully formed conceptions which are known prior to any experience of the extramental world. Nor does it mean that they are mental constructs fabricated by the intellect out of its own "stuff." Self-evident propositions are immediate in the sense that they do not depend on prior propositions for evidence, but they do not arise in us as from a vacuum. They proceed from prior sensory experience by an immediate induction very intimately connected with the abstractive induction of the universal concept from its sensible manifestations.

The truth of a self-evident proposition is seen once the meanings of its terms are known. These meanings are grasped in an abstractive induction of the intelligible object of apprehension from the particular phenomena of the sensibly existing thing. Thus a self-evident proposition is grasped without reference to prior propositions, but only after (and, in a sense, from) sensory experience sufficient to ground an abstractive induction of terms which are in themselves suited to be the extremes of a self-evident proposition.

The self-evident proposition is not simply a report on a factual situation. Yet it is not totally nonempirical. It is, as we have seen, known by way of an immediate induction from sensible data. Nevertheless, it would be misleading to say that it depends directly upon empirical data for its verification. Let us grant that the whole would not be greater than its parts if all wholes were not in fact greater than their parts. Nevertheless, given the meanings of *whole* and *part*, wholes must be greater than their parts, as long as they are at all. Thus, if one understands the meanings of the terms in the proposition *The whole is greater than any of its parts*, one immediately assents to this quite apart from any and all actual instances of sensibly existing wholes and parts. The motive for assent is, as it were, built into the content of the proposition itself. We might say that a self-evident proposition has an intrinsic intelligibility which necessarily constrains the mind to assent. I do not have to have a concrete whole and its parts in front of me to know the whole is greater than its parts. It is enough to know the meanings of whole and part. Yet I would never know those meanings if I never knew any concrete whole and its parts. And, of course, there is no reality at all to whole or part except in so far as there are (at least possibly) concretely existing wholes and parts. For this reason, self-evident propositions are inductively attained only from an experience of singulars, and they make sense only in so far as they bear upon singulars. Yet they are only materially dependent upon experience for their verification. Formally they are verified in their own necessary intrinsic intelligibility, which precludes the possibility even of conceiving the opposite.

VI. Mediate Induction

The real world is a world of established regularity. It makes no sense to conceive of it as though it were fundamentally chaotic, with no explanation for its phenomena save chance alone. The universe of things is, as a whole, an ordered universe. Moreover, each of the beings in it is fundamentally intelligible with a determinate nature, which is what it is because of its ordination to a determinate end. Chance events occur, but when they do occur they are the results of intercepting lines of final causality. Chance itself is inexplicable except in terms of an underlying order in nature.

The existents in the real world are singular things. They are incommunicably distinct, as things, from one another. However, several individuals can fall under the same species, and, along with members of other species, under successively more universal genera and finally under an ultimate genus, which is a determinate mode of being itself. Each determinate type of being, generic or specific, is constituted intrinsically by intelligible principles proper to the essence in question. These essential principles are multiplied from individual to individual, while remaining, of course, intelligibly identical. These principles are the cause of properties in things. These properties are, like the principles which cause them, numerically distinct from individual to individual while remaining intelligibly identical. The connection between the principles proper to one (one in the sense of being intelligibly one) essence and a property necessarily consequent upon them can be expressed in a universal proposition. This universal proposition can be spoken of as a law, since it enunciates an infallible connection between a nature and one of its attributes which cannot not-be realized in any singular instance of that nature.

We have seen that some universal propositions are known immediately upon the understanding of the terms in the proposition itself. These are the self-evident propositions, which are acquired by way of immediate induction. In some cases, however, very often because the subject in question cannot be adequately defined to begin with, it is not self-evident (at least to us) that a universal law is true. This does not mean, necessarily, that it is beyond our reach. Precisely because of the underlying order in reality, whenever a given attribute is found regularly with a given subject, despite varying circumstances, there is sound reason to presume that there is a significant connection between the attribute and that subject. Thus, given a sufficiently discriminative enumeration of singular instances of a given subject with a given attribute, it is valid to conclude that the attribute belongs necessarily to the subject. Here we have induction, namely, the passage from the singular to the universal. Moreover, we have a special kind of induction, which differs significantly from the immediate induction of self-evident propositions. This is a case of inductive argumentation. It is a mediate induction proceeding from a sufficient enumeration of singulars to a universal truth which is particularized in each of the singulars.

We must be careful to note very clearly the manner in which mediate induction differs from the immediate induction of self-evident propositions. In immediate induction there is no question of any movement from premises to a conclusion. The self-evident proposition is in no way a conclusion. It does not look for its evidence to any prior propositions, but is known to be true in the light of the meanings of its terms. It is a universal law which is immediately seen to be true. This is not the case with the proposition achieved by way of mediate induction. This proposition has the characteristics of a conclusion. Since it is a universal proposition, it cannot be factually evident (though each of the particulars which manifest it may be a factually evident proposition). Nor is it a self-evident proposition, since its truth is not immediately evident in the light of its terms. It is a proposition which looks beyond this or that fact, as well as beyond itself, for its evidence. It is seen to be true only in the light of a sufficient enumeration of singular instances of itself. The mind moves legitimately from the knowledge that an attribute belongs to a given subject in this case, and in that case, and in a sufficient number of cases, to the knowledge that the attribute belongs universally to every instance of this subject. The principle of manifestation in mediate induction is not an insight into the intrinsically necessary intelligibility of a universal law. It is the sufficiency of an enumeration of singular instances of this universal law. The evidence for the inductive conclusion is the enumeration itself. If it happens that an enumeration occasions an insight into the intrinsically necessary intelligibility of a universal proposition so that it is seen to be necessarily true in the light of its terms, the proposition ceases to be an inductive conclusion and takes on the character of a proposition which is self-evident to us. At this point the enumeration itself ceases to be of logical significance as a principle of manifestation, though it has served materially to ground the intuition of meaning which is now of prime logical significance. When argumentative induction does not give way to an immediate induction of a self-evident proposition, the sufficient enumeration of singulars remains the principle of manifestation which supplies the evidence for the universal law which is the inductive conclusion.

We must be as careful to distinguish the inductive conclusion from the deductive conclusion as we have been to distinguish the inductive conclusion from the self-evident proposition. The inductive and deductive conclusions are alike only in so far as each is a mediate proposition, that is, a proposition reached only as the term of rational discourse. Beyond this, however, they differ radically. The deductive conclusion depends upon a middle term simultaneously identified with each of its extremes and thereby manifesting their identity (or if the middle functions as a principle of separation, their nonidentity). The inductive conclusion looks to an enumeration of singulars, each of which is a particular instance of it, to manifest the connection between its extremes. Later we shall note that the inductive enumeration of singulars lacks the rigor of the deductive middle term, so that, in practice at least, the inductive conclusion is ordinarily dialectical at best. For the moment it is important simply to stress the radical difference between the inductive argument and the syllogism. The two are irreducibly distinct forms of argument precisely because they rest upon distinct logical principles. The basic principle of syllogistic procedure is the principle of triple identity: two terms each of which is identified with the same third term are identified with one another. On the basis of this self-evident logical principle, the categorical syllogism proceeds to show the identity between the extremes of its conclusion through a middle term with which these extremes are both identified, e.g.:

> Every animal is mortal.
> But every man is animal.
> Therefore, every man is mortal.

The basic principle of mediate induction, however, is not the principle of triple identity but rather what we can call the principle of sufficient enumeration: what can be said of many sufficiently enumerated instances of a universal subject can be said of that universal subject itself. On the basis of this self-evident logical principle, the inductive argument manifests the identity between the extremes of a universal proposition by enumerating singular instances of this universal proposition sufficient to warrant the intellectual "leap" from the singular to the universal, e.g.:

> This man is mortal; and that man
>   is mortal; and this other man
>   is mortal; etc.
> Therefore, every man is mortal.

In so far as both deductive and inductive argument "work" in virtue of distinct logical principles, each involves a distinct mode of inference. The two may be interrelated, as, for example, whenever the conclusion of an inductive argumentation is used as a deductive major premise. Nevertheless, they remain irreducibly distinct forms of argumentation, each with its own distinctive role to play in the life of learning.

VII. The Problem of the Inductive Enumeration

We have seen that in inductive argumentation an enumeration of singulars assumes the role played by the middle term in the syllogism. However, the middle term in the syllogism, at least from the point of view of formal logic, poses much less of a problem than the inductive enumeration of singulars. Given a term which is fully extended at least once, and which is identified with both the major and minor terms (or, in the negative syllogism, identified with one and separated from the other), the syllogistic conclusion follows necessarily from the premises. Furthermore, if those premises are true, the conclusion follows necessarily as true. There is, from the formal point of view, little difficulty with the middle term in the syllogism. Either the term proposed as the middle term is a middle, or it is not. It is not difficult to decide this issue. However, in the case of argumentative induction there are special problems with the enumeration of singulars which manifests the connection between the extremes of the inductive conclusion. We have said that a *sufficient* enumeration of singulars is required for an inductive conclusion. This seems to allow for the possibility of an enumeration which is not sufficient. This is precisely the problem. What constitutes sufficient enumeration? No a priori rule can be given which will cover all cases. The best we can say is that an inductive enumeration is sufficient whenever it rules out the possibility of any singular which would contradict the inductive conclusion. The trouble with this, of course, is that we are saying no more than that an enumeration is inductively sufficient whenever that enumeration is sufficient for induction. Admittedly, this is redundant, and yet it is the best that we can do. Given the fundamental order and regularity of the universe, there is a point at which the human intellect can rise from the singular instances of a universal proposition to the universal proposition itself precisely because there is no reason sufficient to explain the recurrence of these singulars save the truth of that universal law. This is the point of sufficient enumeration. It cannot be numerically located. It is not enough simply to multiply singular cases in order to acquire evidence sufficient for induction. The observations upon which induction rests must be discriminative rather than numerous. For example, it is not enough simply to observe that a given attribute is found a great many times with a given subject to conclude that it belongs universally to that subject. Suppose it happens in each of the many singular cases observed that one and the same accidental characteristic happens to belong to the subject along with the attribute in question. Under the circumstances it is conceivable that the attribute follows upon the accident and not upon the subject in question. No matter how the observations of this subject are multiplied, no conclusion is seriously warranted until the subject is observed without the accidental characteristic. If the subject, observed apart from the accidental characteristic, is seen to have the attribute in question, it is clear that the attribute does not follow from the accident and may well belong necessarily to the subject. One observation of the subject without the variable condition of the accident is worth more than any number of observations with the accidental variable, so long as there has been some observation of the subject with the accident. Only when the singulars observed have been discriminatively evaluated so as to take note of all the variables can they be deemed sufficient to yield an inductive conclusion. In most cases there must be a good number of singulars observed before a conclusion follows. Yet, in no case can we state a priori just how many this will be. In every case, however, we must insist that more than a *numerical* multiplication of singular instances is necessary in order to ensure sufficiency.

We have spoken of sufficient enumeration, and we have indicated that it is difficult to explain the point at which enumeration is sufficient for induction. There is, however, one instance of enumeration which must be sufficient. This is the case of complete enumeration. Inductive enumeration is complete whenever all the subjective parts of a universal are included in the enumeration. Clearly if B can be said of *all* the instances of A, then every A is B. However, there are several points which must be brought out here in order to explain the possibility of complete enumeration on the one hand and the dangers involved in misunderstanding it on the other.

First of all, we have consistently spoken only of the enumeration *of singulars*. Lest we be misled, let us hasten to note that the inductive movement can be correctly described as proceeding from the less universal to the more universal. It is not the case that in every instance of induction the movement is precisely from singulars to a universal. This is the primary instance of induction, and the one in the light of which the others are best understood. However, it is possible to move to a conclusion about a generic subject from a sufficient experience of the various species of this genus. It is impossible to observe *all the singulars* falling under a universal, since there are an infinite number of singular inferiors possible for any concept. But it is possible to know *all the species* falling under a genus. It is possible to know that B belongs to this species of A and to that species of A, and to each and every species of A, and from this complete enumeration it is legitimate to conclude that every A is B. There can be no question here of insufficient enumeration so long as the enumeration is exhaustive. Yet, when it is exhaustive, do we have an advance in knowledge from the enumeration to the conclusion? If X, Y, and Z are the only species of the genus A, and if it is known that X is B and Y is B and Z is B, it seems that there is no advance in knowledge expressed in the proposition Every A is B. Since X and Y and Z taken together are extensively equivalent to A, it seems that to know that X is B and Y is B and Z is B is already to know that every A is B. If there is no advance in knowledge here — i.e., if Every A is B is not different from X and Y and Z are B — then there is no mediate induction. However, it is not necessary to deny an advance in knowledge in Every A is B over X and Y and Z are B. Only in a nominalistic scheme of things, in which the so-called universal is no more than a collection of its so-called subjective parts, must Every A is B be an alternate expression of X and Y and Z are B (when X, Y, and Z exhaust the genus A). In a realist scheme A can be understood as a legitimate generic aspect of things realized in each of its inferiors, but known apart from its specific differences and individuating characteristics. In the proposition *Every A is B* the predicate B is affirmed determinately of the subject A, and this subject represents a legitimate abstraction from the species X, Y, and Z. A is not simply the collection of X and Y and Z. A is a nature which can be considered absolutely or, in other words, for what it is in itself. To know that every A is B is to know that B belongs to A taken absolutely, and through A to each of the inferiors participating in A. So long as a universal subject is not understood simply as the collection of its extensive parts (that is, as a strictly enumerative universal) but rather as one intelligible object (related to its extensive parts as genus to species), there is no reason why a complete inductive enumeration cannot be sufficient for a truly inductive conclusion.

Complete enumeration is always sufficient enumeration. But, as we have seen, complete enumeration is not always possible. Whenever the induction proceeds from singulars in the strict sense, complete enumeration is never possible. Then we are brought face to face with the problem we have already considered: when is the enumeration of singulars (admittedly incomplete) sufficient to warrant the jump from the singulars to the universal conclusion? The best we have been able to do is to suggest that this occurs when the enumeration is quantitatively and qualitatively complete enough so that the possibility of an exception to the universal is ruled out. This occurs when the regularity in the presence of the predicate in a given type of subject is seen as inexplicable unless the predicate belongs necessarily to each and every instance of the subject in question. A conclusion such as this is legitimate only when every conceivably significant factor is checked out. So long as something accidental to the essence of the subject might possibly be suggested as the reason for the predicate in question, it is illegitimate to affirm that the predicate belongs universally to the subject. Under the circumstances it is clear that sufficient enumeration — save in the case of complete enumeration — is extremely difficult. This is at least true for inductive conclusions generated as certainly true. However, this is not the case for probable conclusions to inductive argumentation. Any significant enumeration of singulars generates some degree of probability for the universal that these singulars seem to represent. The more complete the enumeration, the more probable the conclusion. However, so long as there is any possibility of an accidental explanation for the seemingly necessary identity of an attribute with a given universal subject, the proposition expressing this identity is only probable. Because of the extreme difficulty in achieving an enumeration sufficient for certitude, most inductive argumentations yield nothing better than a probable conclusion, though perhaps a conclusion with a very high degree of probability.

Exercise XX

1. Define: induction (generally taken), abstractive induction, immediate induction of principles, mediate induction of conclusions, self-evident proposition, axiom, thesis, inductive enumeration, sufficient enumeration, complete enumeration.
2. What is there about deduction that rules out the possibility that the evidence for all propositions might be deductive?
3. What is the general notion of induction? How is this realized in each of the three types of induction treated in this chapter?
4. What is there about demonstration which necessitates the existence of self-evident propositions?
5. Explain each of these statements:
    5.1. The self-evident proposition is a premise, but not every premise is self-evident.
    5.2. The self-evident proposition is immediate, but not every immediate proposition is self-evident.
    5.3. The self-evident proposition is necessary, but not every necessary proposition is self-evident.
    5.4. The self-evident proposition is indemonstrable, but not every indemonstrable proposition is self-evident.
6. What is the difference between a proposition self-evident only in itself and a proposition self-evident to us? Illustrate.
7. What is the difference between an axiom and a thesis? What are the different roles each has to play in scientific argumentation? Illustrate.
8. Is there any distinction to be made between self-evident propositions and propositions involving a mode of perseity?
9. In what way do self-evident propositions depend upon the data of sense experience? How are self-evident propositions verified?
10. What is wrong with a description of the self-evident proposition which identifies it as a proposition clearly and easily understood?
11. What is there about reality which makes mediate induction possible?
12. What precisely is the difference between the evidence for a self-evident proposition and an inductive conclusion?
13. What precisely is the difference between the evidence for an inductive conclusion and the evidence for a syllogistic conclusion?
14. Why are induction and deduction irreducibly different modes of inference?

INDUCTION

15. Explain: in the inductive argumentation complete enumeration is always sufficient, but sufficient enumeration is not always complete.

16. Under what circumstances is complete enumeration a possibility? Illustrate.

17. What does an inductive conclusion add to a complete enumeration so that it can be considered truly an advance in knowledge and thus a conclusion?

18. Sufficient enumeration is ordinarily sufficient at best for a probable conclusion. Explain.

19. The following rules[2] are the famous canons for induction proposed by John Stuart Mill, a nineteenth-century logician who is best known for his discussion of the inductive method. Discuss the value of each one, especially with an eye to the discriminative character of the inductive enumeration.

   19.1. If two or more instances of the phenomenon under investigation have only one circumstance in common, the circumstance in which alone all the instances agree is the cause (or effect) of the given phenomenon. (Method of Agreement)

   19.2. If an instance in which the phenomenon under investigation occurs, and an instance in which it does not occur, have every circumstance in common save one, that one occurring only in the former; the circumstance in which alone the two instances differ is the effect, or the cause, or an indispensable part of the cause, of the phenomenon. (Method of Difference)

   19.3. If two or more instances in which the phenomenon occurs have only one circumstance in common, while two or more instances in which it does not occur have nothing in common save the absence of that circumstance, the circumstance in which alone the two sets of instances differ is the effect, or the cause, or an indispensable part of the cause, of the phenomenon. (Joint Method of Agreement and Difference)

   19.4. Subduct from any phenomenon such part as is known by previous inductions to be the effect of certain antecedents, and the residue of the phenomenon is the effect of the remaining antecedents. (Method of Residues)

   19.5. Whatever phenomenon varies in any manner whenever another phenomenon varies in some particular manner, is either a cause or an effect of that phenomenon, or is connected with it through some fact of causation. (Method of Concomitant Variations)

20. Evaluate the following attempts at induction. Add to these other examples of induction taken from your textbooks and reference works in other classes.

---

[2] John Stuart Mill, *System of Logic*, Bk. III, Ch. VIII (New York: Longmans, Green and Company, 1930), pp. 255, 256, 259, 260, 263.

20.1. Since the scalene triangle has three interior angles equal to two right angles, and since the same is true of the isosceles triangle and the equilateral triangle, it follows that every triangle has three interior angles equal to two right angles.

20.2. Since scotch and soda, rye and soda, brandy and soda, bourbon and soda, and gin and soda have all been observed to make a man drunk, it is clear that soda makes a man drunk.

20.3. As long as a large number of observations have been made of liquid boiling at 100 degrees centigrade, it follows that all liquid boils at 100 degrees centigrade.

20.4. Since the senses of sight, hearing, taste, smell, and touch each demand a corporeal organ, it follows that the senses as such demand a corporeal organ.

20.5. Since all the snow that we have ever seen, or that anyone has ever seen, has been white, we can conclude that all snow is naturally white.

20.6. Since samples of all known metals have been observed to conduct electricity, it follows that every metal conducts electricity.

20.7. Several metals, for example, lithium, potassium, and sodium, are lighter than water. Therefore, it is safe to conclude that all metals are lighter than water.

20.8. Since the immediate inferior of species, specific difference, specific property, and accident is in each case a singular, it follows that the immediate inferior for every predicable is a singular.

20.9. Since every healthy adult human being speaks, it follows that the ability to speak belongs by nature to man.

20.10. Men die, brutes die, and plants die; hence, all living things die.

CHAPTER XXI

Fallacious Argumentation

I. The Notion of the Fallacies

Any logically erroneous procedure in rational discourse is spoken of as a fallacy. Thus, any logically defective argumentation is said to be fallacious. This is the case whether the defect is on the side of form or on the side of matter. Strictly speaking, an argumentation which is designedly fallacious and intended to deceive is said to be sophistical. Sophistry can be defined as the art of deceptive argumentation. Sophistry aims at the appearance of truth, but has no commitment to the truth itself.

In a sense we have already considered the fallacies. We have investigated the rules for proper logical procedure and have faced the possibility that each might be violated. The violation of any one of the rules for sound discourse is a fallacy. Thus, we may speak of the fallacy of four terms, the fallacy of the unextended middle term, the fallacy of the illicit major or minor term, and the fallacy of demonstrating from merely probable premises. We could add to our list of fallacies by naming the violation of each of the rules we have already discussed.

There is no limit to the ways in which the rules for sound discourse can be violated. There is only one legitimate construction for any given logical form, but there is an indefinite number of ways in which a defect can be introduced into this construction. We have not been able to suggest them all, nor do we intend to do so in this chapter. However, there are some rather common fallacies which are traditionally given special consideration. We shall take them up rather briefly in this chapter. A direct treatment of some fallacies helps, by contrast, to heighten our appreciation for licit discourse. A special study of these common fallacies puts up our guard against them. The fallacies which are traditionally considered fall into two classes. The first is generally grounded upon some defect or difficulty on the side of verbal expression. These are the fallacies of language. The second is rooted in something other than a semantic confusion. These are the *fallacies beyond language*.

II. The Fallacies of Language

A. *The fallacy of equivocation*

The fallacy of equivocation is the use of identical oral or written terms to stand for diverse meanings in the same discourse. This includes the use of strictly equivocal terms and the illegitimate use of analogous terms. It also includes the use of terms so vague as to be practically meaningless. We might even take it to include the use of terms which, though standing for identical meanings, differ in supposition from proposition to proposition within an argumentation. Terms are *conventional* signs, and it is inevitable that they will be used in different ways at different times. Logical discourse requires a rigorous determination of the meanings of the terms involved and a uniformity in the use of these terms in the propositions of the discourse. Otherwise, the argument will lack integrity, for it leaves in doubt the status of the terms, which are the fundamental elements into which any discourse can be finally resolved.

Examples of the fallacy of equivocation are so common that it is not necessary to supply them here. However, the student should refer to Chapter X to review the discussion on the unity of signification and supposition needed for valid discourse, and to reconsider the examples offered there of the fallacy of equivocation.

B. *The fallacy of ambiguity in expression*

The fallacy of ambiguity in expression is the use in discourse of badly expressed phrases which readily admit of misinterpretation. This differs from the fallacy of equivocation in that here the confusion is not in the terms themselves but in their syntactical arrangement. The fallacy of ambiguity in expression is sometimes called amphiboly.

The following examples of amphiboly should make it very clear that this fallacy is inconsistent with clear and rigorous expression in logical discourse:

1. "Logic is necessary for metaphysics because it is scientific."
   (What precisely is scientific — logic or metaphysics?)
2. "The free world the Communist world will survive."
   (Who precisely will survive whom?)
3. "Kathleen loves Susan because she is her mother."
   (Who is mother of whom?)

C. The fallacies of composition and division

The fallacy of composition consists in combining words which should be understood separately, and the fallacy of division consists in separating words which must be understood together.

Consider the proposition: "The only nonessential predicables are properties and accidents." This makes sense only if "properties" and "accidents" are understood separately. It is enough for a predicable to be a property (or, if not a property, an accident) for it to be a nonessential predicable. To misconstrue the proposition and understand it as demanding that a nonessential predicable be simultaneously a property and an accident is to commit the fallacy of composition.

Consider this other proposition: "The major premise and minor premise are the antecedent in the syllogism." This proposition is true only if "major premise" and "minor premise" are taken together as one logical subject, for they represent the antecedent of the syllogism only in so far as they are understood as one pair. Neither the major premise nor the minor premise alone is the antecedent of the syllogism. To take them separately as though the predicate belongs divisively to each is to commit the fallacy of division.

D. The fallacy of accent

The fallacy of accent is the misuse of an accent in order to place undue emphasis upon some element of a proposition and thereby deceive. The addition of a special accent to the ordinary proposition can alter its meaning or add something to it. Consequently, one must be careful in accentuating propositions and in interpretating propositions involving any unusual accentuation.

To illustrate how significantly an accent can modify the meaning of a proposition, consider these two examples:

1. "Logic is not a science *of real things*."
2. "Logic is not a *science* of real things."

The first example, in which "of real things" is accented, suggests that logic is a science, but that it is not a science of real things. This is true. The second, in which "science" is accented, suggests that logic may be of the real, concerned with real things, but that it is not a science. This is false. Other examples of this fallacy include:

1. "*Some* loyal Americans hate Communism."
    (Though an *I* proposition does not, on its own merits as an *I* proposition, imply its subcontrary, this proposition, because of its accent, implies "Some loyal Americans do not hate Communism.")

2. "Aristotle was a good *logician*."
   (The accent added here makes this proposition imply that Aristotle was not good at things other than logic. Without the accent, this proposition leaves this an open question.)

E. *The fallacy of similar word construction*

This is the fallacy involved in thinking that any similarity in verbal expression is a sign of a similarity in meaning. Grammatical expression frequently reflects the logical situation rather accurately, but this is not always the case. Language is to an extent arbitrary, but logic is not.

To infer that "inflammable" and "invaluable" mean "not flammable" and "not valuable" just because "insincere" and "invalid" mean (at least) "not sincere" and "not valid" is to commit the fallacy of similar word construction. Actually, "inflammable" and "flammable" mean the same thing, while the "in" in "invaluable" intensifies the adjective instead of negating it.

These five fallacies do not represent all the fallacies of language, although they represent the most significant. They are important in themselves, since they make themselves felt frequently and subtly. They are also important as a serious reminder that all fallacies of language stand in the way of profitable scientific (as well as nonscientific) discussion. Although there are many serious problems that cannot be solved simply by knowing what the terms involved mean, we can never begin to work on their solution unless we first eliminate completely whatever semantic confusions originally exist with respect to them.

III. Fallacies Beyond Language

A. *The fallacy of accident*

The fallacy of accident consists in mistaking the accidental (or non-essential) for the essential. Some things are affirmed or denied of a subject in virtue of something essential to it. Other things are affirmed or denied of a subject in virtue of something nonessential to it. To confuse these two is to commit the fallacy of accident.

The following examples of this fallacy should be sufficient to indicate the frequency and serious consequences of its use:

1. An argument against a man's right to life, liberty, and the pursuit of happiness on the grounds of the color of his skin.
   (The right to life, liberty, and the pursuit of happiness is a right consequent upon human nature as such and not upon an accident of human nature such as the color of a man's skin.)

2. An argument for the condemnation of all books because some books are obscene.
   (An obscene book deserves condemnation not in virtue of its essential character as a book, but rather in virtue of its accidental state of obscenity.)
3. An argument designed to show that a man is not a good artist because he is not a good man.
   (The question of whether or not any man is a good artist is quite accidental to the question of whether or not he is a good man. A man is good as a man if his will is rectified. The rectification of the will is compatible with the absence of artistic talent, while artistic talent is compatible with an unrectified will.)

B. *The fallacy of mistaking the qualified for the absolute*

The fallacy of mistaking the qualified for the absolute consists in moving from something true only relatively or in a special situation to the unjustified assumption that this is true absolutely and universally. This fallacy includes the unwarranted jump from the truth of an I or O proposition to the truth of the corresponding A or E.

The following examples illustrate this fallacy:

1. An argument designed to show that any man can do and say whatever he chooses to do and say on the grounds that a man is by nature free.
   (Man is free in only a qualified sense. There are limits to human freedom which are ignored in this argument.)
2. An argument against capital punishment based on the commandment of God: Thou shalt not kill.
   (This commandment is not to be taken absolutely. There is the possibility that one man may in justice take the life of another, e.g., in self-defense. This argument falsely presumes that the commandment in question applies without qualification.)
3. An argument that a given pan of water will in fact boil at 100 degrees centigrade because all water boils at 100 degrees centigrade.
   (It is true that all water boils at 100 degrees centigrade, but only at sea level and only without impurities. This pan of water may not be at sea level and may contain many impurities.)

C. *The fallacy of arguing beside the point.*

The fallacy of arguing beside the point is the practice of ignoring the proper issue in an attempt to make a point by way of something extrinsic to it. Next to the fallacy of equivocation, this is the most common and persistent fallacy. As a matter of fact, equivocation and arguing beside the point can go hand in hand. One reason for ignoring an issue is to miss it completely through a misunderstanding of the terms used in it. There are many examples of the fallacy of arguing beside the point. We shall consider several of them. The species of this fallacy are best known by their Latin names, and we shall use these names to identify them.

a) *Argumentum ad hominem* (argument aimed at the man):

The *argumentum ad hominem* confuses the person of one's opponent in debate with the real issue of the debate. Whenever one departs from the true issue to attack one's opponent, one uses the *argumentum ad hominem*. This is the method used, for example, by a man who questions the political theory of another on the grounds that the other has been an unsuccessful businessman or has had a criminal past. Again, it is not relevant to argue that a man's philosophical stand on a moral issue is false simply because the man himself is immoral, or because he was previously shown to be wrong on a question in nuclear physics, or because he once had a friend who was a Communist. It is never legitimate to attempt to make a point by discrediting the person of an opponent, unless the opponent's argument is in some way based upon his personal integrity.

b) *Argumentum ad populum* (argument aimed at the people):

The *argumentum ad populum* is an appeal to popular passion, school spirit, racial enthusiasm — that is, to some kind of partisan prejudice — in order to make a point which has nothing to do with passion or prejudice. It may be quite easy, regardless of the facts, to convince a Frenchman that a war against Germany is justified, or a Southerner that racial integration should be indefinitely put off, or a Marquete University student that Marquette's basketball team is the country's finest. However, so long as these arguments rely on some prejudice in their hearers and lack solid premises which are to the point, they are simply instances of an *argumentum ad populum*.

c) *Argumentum ad baculum* (argument aimed at the stick):

The *argumentum ad baculum* is an attempt to win acceptance of a point by the threat of force rather than by backing it up with solid premises establishing a conclusion. A logic teacher who argues that students should study logic in order not to fail the course, or a practical politician who solicits votes by threatening to do violence to those who refuse to vote for him, or an executive who wins approval for his policies by threatening subordinates with dismissal — all these employ the *argumentum ad baculum*.

d) *Argumentum ad verecundiam* (argument aimed at respect):

The *argumentum ad verecundiam* is an appeal to misplaced authority in an attempt to establish a point. It is a legitimate dialectical method to argue from the opinion of an expert in a given field to a conclusion in that field. However, it is not legitimate to appeal to an authority in one field to establish a point in another. To do this is to employ the *argumentum ad verecundiam*. Examples of this fallacy are thrust at us in our magazines and newspapers, and over our radios and television sets. Movie actresses give advice on auto insurance, baseball players on shaving equipment, and jazz musicians on wall-to-wall carpeting. There are, more unfortunately, examples of this fallacy found even in academic circles, where experts in one field frequently presume upon their authority in this field in order to make a point in some distinctly different academic area. Thus an expert psychologist may expect a hearing on philosophy, and an expert physicist on theology, or an expert mathematician on law.

e) *Argumentum ad misericordiam* (argument aimed at pity):

The *argumentum ad misericordiam* is an appeal to a man's sympathy rather than to his reason in order to make a point. This fallacy is very similar to the *argumentum ad populum*. Both depend upon the fact that very frequently the "heart leads the head." The timeworn picture of the eloquent lawyer pleading for the life and liberty of a widow (perhaps, in fact, a widow by her own hand) accused of a crime by pointing to her poverty, her grief, her defenselessness, her femininity, etc., is the picture of a man employing an *argumentum ad misericordiam*. The widow's life and liberty should depend on guilt or innocence, and neither guilt nor innocence is evidenced by grief or poverty or the like. A student who argues for a higher grade lest he incur the displeasure of his parents, as well as the youngster who answers a parental accusation with a sob, also employ the *argumentum ad misericordiam*.

D. *The fallacy of begging the question*

The fallacy of begging the question consists in assuming in the premises of an argument the conclusion which ought to be proved. It is true that the premises in a valid argumentation cannot be true without the conclusion being true: this is the very essence of validity. Yet the truth of a premise cannot rest on the truth of the conclusion. The reverse of this must be the case. So long as a premise is simply a restatement of the conclusion, or is understood only in so far as it follows from the conclusion, or is in any way dependent upon the conclusion for its truth, that premise cannot function validly as a principle for establishing that conclusion. This fallacy is especially tempting to one who is avidly concerned to defend a proposition. A man sometimes so intently wants his position accepted that he assumes it in practice and uses it in its own defense.

The following examples are illustrations of this fallacy:
1. An argument for the immortality of the soul from the fact that the soul cannot die.
2. An argument calculated to show that a dog can think because dogs are intelligent.
3. An argument to prove that at least one premise in a categorical syllogism must be affirmative since they cannot both be negative.

Perhaps the most evident instance of begging the question is found in an argument which involves the so-called vicious circle. One argues in a vicious circle whenever he attempts to prove a first proposition by a second proposition, and then that second proposition by the first. The following examples illustrate that particular type of begging the question represented by the strictly circular argument:
1. An attempt to establish credence in a man's political position on the basis of his reputation as a political theoretician — coupled with an attempt to establish his reputation as a theoretician precisely on the basis of this political position.
2. An attempt to prove that subcontraries can't be false together precisely because contraries can't be true together — coupled with an attempt to prove that contraries can't be true together because subcontraries can't be false together.

We have spoken about the problem of begging the question earlier in the book. We saw that a categorical syllogism with an enumeratively universal major premise begs the question, for its conclusion expresses something which is already expressed in its major premise.

E. *The fallacy of the false cause*
There is frequently the temptation to argue that one thing happens *because* of another because it happens *after* that other, the *post hoc, ergo propter hoc* type of argument. This is to argue fallaciously by attributing a false cause to an event or fact. It may be true that something which happens after something else is caused by it. Yet temporal

sequence is no guarantee of causal sequence. With nothing but temporal sequence as evidence, no conclusion (save that of possibility) is warranted.

The following examples illustrate this fallacy:

1. **An argument that Hoover caused the great depression because it came upon the country shortly after Hoover took office as president.**
2. **An argument designed to show that a given medicine has cured a certain illness because a person suffering from the illness was cured after taking this medicine.**

A slight variation of this fallacy is found in an argument which confuses an accompanying attribute with a cause and in an argument which mistakes an occasion for a cause. I cannot argue that my breathing causes me to think just because I must breathe at the same time that I think. Nor can I say that a party causes me to gossip, although it does offer the occasion for my doing so.

F. *The fallacy of the consequent*

With the exception of reciprocal conditional propositions, the truth of a consequent does not imply the truth of its antecedent, nor does the falsity of an antecedent imply the falsity of its consequent. To proceed as though they do this is to commit the fallacy of the consequent. We have said enough on this in our chapter on hypothetical argumentation.

G. *The fallacy of the complex question*

To pose a question calling for a "yes" or "no" answer in such a way that any answer given to it implies more than a simple affirmation or denial is to commit the fallacy of the complex question. It is unfair to demand a simple "yes" or "no" answer to questions such as this: "Was his effort an honest and successful one?" The simple negative answer seems to imply that the effort was neither honest nor successful. A simple affirmation seems to imply it was both. If the effort was honest but not successful, neither the "no" nor the "yes" answer suffices. Each seems to imply more than the truth. A distinction should be made, to allow for the possibility that his effort was truly honest but not successful. It is also unfair to ask a man whether or not he is still as dishonest as ever, or still stupidly engaged in some enterprise, or not yet over a bad habit. Both a "yes" and a "no" answer will imply at least something unpleasant for the man who replies. Perhaps he has never been dishonest. Perhaps his enterprise was never stupid. Perhaps the habit spoken about is a good habit. In these cases neither a "yes" nor a "no" answer can do justice to the facts. Each implies something which is not true.

Exercise XXI

1. Define: fallacy, sophistry, fallacy of equivocation, fallacy of ambiguity in expression (amphiboly), fallacy of composition, fallacy of division, fallacy of accent, fallacy of similar word construction, fallacy of accident, fallacy of mistaking the qualified for the absolute, fallacy of arguing beside the point, *argumentum ad hominem*, *argumentum ad populum*, *argumentum ad baculum*, *argumentum ad verecundiam*, *argumentum ad misericordiam*, fallacy of begging the question, argument by way of a vicious circle, fallacy of the false cause, fallacy of the consequent, fallacy of the complex question.
2. Distinguish strictly between fallacious and sophistical argumentation.
3. Relate the chapter on fallacies to the previous chapters in the book.
4. Distinguish between fallacies of language and fallacies beyond language.
5. Examine each of the following examples in order to uncover the fallacy (or fallacies) involved. In many cases the examples represent argumentations which have been actually and seriously attempted.
    5.1. A Russian scientist has argued that successful flights of earth satellites and rockets cast doubt on the existence of God and angels. The reason he has given is that rockets and satellites in flight have encountered no angels nor discovered a supreme being.
    5.2. A well-known educator suggests as an argument for coeducational schools the fact that the family is naturally coeducational.
    5.3. A segregationist has argued for "all-white" schools in the South on the principle that "mixed" schools violate the individual's right to freedom of association, that is, his right to associate with whom he pleases.
    5.4. A New England "Yankee," who has always suspected as much, takes pleasure in the admission on the part of a Hibernian that the Irish are thick around Boston.
    5.5. A nationally known economist has argued that the progressive income tax is Marxian because it takes a disproportionately high amount from high incomes.
    5.6. A poor philosopher argues that since we cannot prove our ability to know anything with certitude we must either blindly assume that we can or admit to skepticism.
    5.7. A well-known Midwestern brewery suggests in its advertising that its beer is of a superior quality because it is brewed and bottled *only* in Wisconsin.
    5.8. A man long on an interest in cats and short on logic argues that one cat has ten lives, because no cat has nine lives, and one cat has one life more than no cat.
    5.9. A phone company has advertised that one phone is not enough in a modern home because a modern home has plenty of phones.

5.10. A mathematician disputes the right of a philosopher, as philosopher, to discuss the nature of mathematics on the grounds that it is the business of the mathematician and not the philosopher to know mathematics.

5.11. An objection is posed against the Thomistic position that there is an interior sense, which can be spoken of as the common sense, in addition to the five exterior senses of sight, hearing, taste, smell, and touch. The objection rests upon the fact that the common is not divided off against the proper. The species of a genus are divided off against one another, but not against the genus itself.

5.12. An American Communist argues for the right to teach Marxian economic theory in our American schools on the grounds that our constitution guarantees freedom of thought and speech to all its citizens.

5.13. This same Communist argues for the integrity of Marxian economic theory on the grounds that it opposes capitalistic oppression and exploitation of the working class.

5.14. A skeptic expresses amazement that a Thomist philosopher claims to be, sometimes at least, absolutely certain in his knowledge of things. The Thomist, he argues, should know that only God has an absolutely certain knowledge of things.

5.15. The ontological argument for the existence of God finds the guarantee of God's existence in the concept of God. Since the concept of God is the concept of a being which cannot not-exist, then, since the concept of God exists, God must exist.

5.16. A pharmaceutical house has advertised a salve by noting that some arthritic patients who have used the salve along with physical therapy have experienced relief shortly afterward.

5.17. Aristotle's claim that logic should be a tool for the other sciences seems to suffer in the face of the fact that for the most part his only logical works are those included in the *Organon*.

5.18. One might be tempted to approve the devastation wrought by the divisions of Hitler, since it is certainly true that the destruction of Hitler's war machine was justified.

5.19. A pagan wonders about the slogan of Christians: "Put Christ back into Christmas." How else can you spell Christmas except that the first syllable be Christ?

5.20. Some modern psychologists question St. Thomas' argument for the spirituality of the human intellect, which is based upon the fact that the human intellect knows an object abstracted from its individuating characteristics. They say that the modern appreciation of the effect of damage to the cerebral cortex upon the thinking process is more sophisticated than that possessed by St. Thomas.

5.21. It is absurd to distinguish between philosophy and the history of philosophy, since philosophy never exists except in the mind of a philosopher, who, as a philosopher, becomes a figure in the history of philosophy.

5.22. Wherever there are a number of things, these things can be counted. Thus, a man with countless blessings must be a man who is completely unblessed.

5.23. If one is a criminal, he is responsible for his actions; and if one is responsible for his actions, he is a man. Therefore, all men are criminals.

5.24. Geometry is rigorously logical and metaphysics is in no way geometrical; thus, metaphysics is in no way logical.

5.25. It is a good thing to relieve the suffering of an innocent person whenever this is possible. Thus, mercy killing is a good thing whenever it is possible.

5.26. Men and women have a right to an equal status in all things since men and women are essentially on a par with one another.

5.27. The most significant philosophical thought in Greece must have found its cause in the teaching of the Sophists, since they immediately preceded the greatest Greek philosophers, namely, Socrates, Plato, and Aristotle.

5.28. It has been argued that the position which the Catholic Church takes on birth control is erroneous precisely because the Catholic clergy is committed to a life of celibacy.

5.29. The president of any university must be an extremely respected man, since the position of university president is an extremely respectable position.

5.30. A skeptic argues that no one can deny that no proposition can be accepted without fear of its opposite, since every proposition admits of contradiction.

6. Carefully examine issues of the daily newspapers, copies of magazines, even your textbooks, for examples of fallacious argumentation. Make a list of these examples, naming the fallacy wherever possible, and indicating precisely what rule of correct logical procedure is violated in each case.

CHAPTER XXII

The Nature of Logic

I. The Purpose of This Chapter

In the strict sense it is not precisely the business of logic to concern itself with the nature of logic. This is the task of that branch of philosophy called epistemology which investigates the nature of knowledge, the various types of knowledge, and the relationships existing between those types. It is not the business of the mathematician, as mathematician, to know the nature of mathematics — but to know the mathematical subject matter. Similarly, it is not the business of the logician, as logician, to know the nature of logic — but rather to know the subject matter of logic, which we shall speak of as the second intention. Nevertheless, it is of value for any scientist, including the mathematician and the logician, to know something about the nature of his science, to be able to evaluate its significance, and to see precisely its limits. Though this is not *properly* within the sphere of the science in question, it is important for the well-being of that science in the scientist in question. The limits of mathematics are not mathematical, but the mathematics of a mathematician who ignores them may be of questionable integrity. So too with logic and the logician. Knowledge about logic is not to be confused with knowledge in logic. Yet the well-being of knowledge in logic (which is the science of logic) suffers in the absence of sound knowledge about logic (which is epistemological).

Some effort was made in Chapter I to introduce the student to the nature of logic. However, the student's inexperience in logic prior to his course in logic makes any extended treatment of its nature impracticable at the beginning of the course. The nature and limits of logic are not easily understood, and they can easily be misunderstood by one who attempts to investigate them without any experience in logic. It is sufficient at the beginning of the course to discuss only generally the nature of logic and then briefly to indicate the divisions of logic which determine the manner in which the course is ordered. At the end of the course, however, the student should have sufficient experience to understand more determinately the fundamental notions concerning the nature of logic and its division into parts.

II. Art and Science

Logic is both an art and a science. In order to see that this is so, let us consider briefly what it is to be an art and what it is to be a science. We can then measure logic up to these notions.

Art and science are alike in that each is an intellectual perfection which is ordered to some properly human activity. They differ in that each is ordered to a different activity. Man is by nature potentially a knower and potentially a maker. His intellect is naturally adapted to confront a reality which exists independently of him and to know that reality as it is, even to know it scientifically in its causes. His intellect is further adapted by nature to direct him in the production of works which are his own "creation,"[1] but which are in one way or another an imitation of the works of nature. These works "created" by man are spoken of as works of art. The basic ordination of the intellect to knowledge on the one hand and to productive activity on the other is sound. However, the intellect can deviate from the straight path and fall into error in both, since the intellect is not completely determined by nature for either scientific knowledge or artistic effort. Fortunately, it can acquire determination in both respects. In fact, it can be habituated badly or well in either case. A bad habit in either respect is an intellectual vice. A good habit, on the other hand, is an intellectual virtue. The intellectual virtue which perfects the intellect in its quest for scientific conclusions is called science. The intellectual virtue which perfects the intellect so that it can direct a man in his artistic effort without error is called art.

Art differs from science, and from every other intellectual habit, in being essentially ordered to the production, the making, of a work. Since the work is the final cause of the art, the truth of any artistic judgment which bears upon the production of the work depends upon a conformity with the character of the work to be produced. The raw materials for any artistic work are given in nature, and the artist imposes an artificial (rather than natural) form upon these materials in order to produce his work of art. Ordinarily the matter for the artistic work is found in the world of physical things which exist outside the intellect. Here the painter finds

---

[1] In the strict sense, creation is a production of something out of nothing. Only God can create in the strict sense. The production we speak of here — the artistic production of men — presupposes something given in nature to which an artificial form can be added.

his canvas and his paints; the sculptor finds his clay or bronze; and the builder finds wood or bricks or steel. In these cases the work of art depends upon an intellectual effort of the mind, but it is produced in a medium which is exterior to the mind. In other cases the matter of the artistic work is found within the intellect itself. In these instances, of course, the matter for the work of art is not material in precisely the way in which paints, clay, and wood are material. However, it is possible for something to exist in the mind, and in the mind to lack an artificial form which it could take on. When we spoke on the four causes in our chapter on definition we described the material cause as an indeterminate but determinable substratum of form. In a general sense this description fits both the matter for a work of art which is exterior to the intellect and that which is interior to the intellect. Those arts which are ordered to a work existing outside the intellect are traditionally spoken of as servile arts. They include fine arts such as painting and sculpture, as well as useful arts like medicine and carpentry. Those arts which are ordered to an interior work produced within the intellect are called liberal arts. We shall see that logic is a liberal art. Other examples of the liberal arts are grammar and music.

Science (in the Aristotelian sense) is a habit of the intellect perfecting it in its quest for knowledge demonstrated through causes. Scientific knowledge, like all knowledge, admits of a distinction between the speculative and the practical. Speculative knowledge is knowledge which is of value for the sake of knowledge. Practical knowledge is knowledge which is of value for the sake of an activity beyond knowing. In the strict sense, speculative knowledge is knowledge of an object which is given in nature and which cannot be produced by man. Being is an example of such an object. Thus, metaphysics, which is the science of being, is an example of a speculative science. Being is an object well worth knowing. But this knowledge is in no way intended to be used in order to make or modify being. It is purely speculative. Strictly speaking, practical knowledge is knowledge of an object which can be produced in some way by man. It is knowledge of an object which is of value precisely in so far as it can be ordered to an activity beyond knowing. The object of medicine is the curable body, and the object of ethics is the moral act. In each case the object is somehow able-to-be-brought-about, and in each case the end is an activity which is different from the activity of knowing. Metaphysics is a speculative science. It represents, in an unequivocal fashion, knowledge for the sake of knowledge. Medicine and ethics are practical sciences. Each is an example of knowledge for the sake of a noncognitive activity.

III. Logic as an Art

It is not difficult to see that logic is an art, and that it is a liberal art. The human intellect is by nature an ability to reason. We spoke about the native capacity of man for logical discourse, and we called this natural logic. Natural logic, however, lacks the determination which would guarantee an orderly, easy, and infallible movement from the known to the unknown. For this the human intellect stands in need of logic. Logic is an intellectual habit which enables the intellect to impose that order upon its objects which will ensure the orderly, easy, and correct movement from previous knowledge to new knowledge. Logic "firms up" the intellect for the construction (within the intellect, of course) of integral discourse. It is a perfection directing the intellect in the production of interior works such as definitions, divisions, syllogisms, and demonstrations. As a perfection of mind directive of productive activity, logic is an art. Because the work produced by logic is an interior work, logic is a liberal art.

We saw that the truth of an artistic judgment is measured by the good of the work to be produced. A work of art is not wholly arbitrary. Just as the final cause of a work in nature determines its matter and form, so the end of an artistic work determines its matter and form. Given the final cause of a house, for example, it is clear that only some materials are fitted for the production of a house and that the ways in which these materials can be put together are determinately limited. On the side of the matter art demands a certain indetermination so that the matter can take on an artificial form. On the side of the artist art demands possession of certain and determinate rules according to which some determination or form can be produced in the matter. This is precisely the reason why logic must be both an art and a science. We shall see that it cannot be the art of sound rational discourse unless it is also the science of logical relations or second intentions. The reason for this, as we shall see, is that an adequate understanding of the rules of rational discourse is identical with a scientific grasp of logical relations or second intentions.

IV. Logic as a Science

Logic as an art is ordered to rational discourse. It is a perfection of mind which enables us to reason well. It is not enough for adequate reasoning or argumentation that a conclusion be correct, not even that it be correct both formally and materially. It must be correct, and, at the same time, reflexively assured. By this we mean that it must be able to be defended. We have seen that no work of art is completely arbitrary. This is especially true of rational discourse. Rational discourse is sound only if it conforms to strict canons which determine its construction. No one can adequately reason unless he can defend himself in terms of these canons. And no one can defend himself in terms of these canons unless he sees the necessity which imposes them upon his discourse. The rules of argumentation are for the most part particular canons of logical procedure, which are not immediately evident but which find their evidence ultimately in self-evident principles in the logical order. Thus, the six rules of valid procedure for any figure of the categorical syllogism are particular rules whose necessity is seen only if it is shown that they follow necessarily from the self-evident principles of triple identity or the separating third. Similarly, the rule which states that subcontraries cannot be false together is adequately appreciated as necessary only in so far as it is resolved into the self-evident logical principle which asserts that propositions which deny one another cannot be true together. Thus, the art of integral discourse is achieved in a knowledge of the rules of argumentation which have been resolved into the first principles of logical procedure. We have seen that the resolution of particular conclusions into self-evident propositions is demonstration or scientific argumentation. Thus, if logic is to be the art of integral discourse, it must be at the same time a scientific discipline. The rules without which logic is not an art can be adequately possessed only if they are known scientifically.

Every science has its proper object and its proper end. We have seen that for logic to be the art that it is, logic must be simultaneously a science. What is the object of logic as a science? What is its end? These questions might be rephrased so as to be more pointed for us. What does the student in logic class study? Why does he study this, and what does he intend with the knowledge he acquires?

The rules of logical procedure are simply canonical statements of those logical relationships which exist between objects in the mind and which govern the ordering of these objects in rational discourse. These logical relationships are technically spoken of as second intentions. Since logic as a science demonstrates the rules of logical procedure, we can say that the object of logic is the second intention. What precisely is a second intention?

We have seen that natures enjoy a twofold existence. They exist first as concretely individualized in real things. They exist secondly in a state of abstraction, i.e., as known in the mind. The comprehensive notes of a nature belong to that nature taken absolutely, i.e., in itself and open to existence both in things and in the mind. Moreover, these notes can be said of the nature as it exists both in the individual and in the mind. Some attributes belong to the nature precisely in virtue of its existence in the real individual. We call these attributes individuating characteristics. Other attributes belong to the nature precisely in virtue of its existence in the mind as known. As an object in the intellect, and only as such an object, a nature can be seen to be related in different ways to different things. These relationships, which an objective concept possesses precisely as known, are the second intentions studied by the logician. They are called *second intentions* because they belong to natures only in so far as they exist in mind as known, whereas *first intentions* are the natures themselves as the direct object of intellectual knowledge. Moreover, second intentions are properties which natures acquire only in that second existence which they enjoy in the intellect as objects of thought, in opposition to the first existence which those natures have in individual real things. *Man*, as an object of direct intellection, is a first intention. The logician is never interested precisely in what it is to be a man. *Man*, absolutely taken, is the concern of the psychologist or anthropologist. However, in the mind *man* takes on relationships which are the concern of the logician. In the mind *man* takes on the relation of species to John, of subject in the proposition *Every man is rational*, of middle term in the syllogism, *Every man is rational; but John is a man; therefore, John is rational*. The property of being a species, the property of being a subject, and the property of being a middle term — these are all second intentions. Each is a relation belonging to a nature precisely in the second existence the nature has as object in the intellect. These are only three of the many second intentions we have studied in logic. Others include the properties of predicate, essential definition, logical division, hypothetical proposition, obversion, syllogism, primary proposition, explanatory demonstration, and induction. There is no need to add to these examples. A complete list would serve as an outline reviewing the whole of our course. Our investigation has been essentially a study of second intentions.

We have said that the art of correct reasoning is simultaneously the science of second intentions. This can be clear to us only if we see an identity between second intentions and the rules of logical procedure, for, as we have seen, the art of logic is one with the scientific grasp of the rules of logical procedure. We have asserted the identity between second intentions and the rules of logical procedure by saying that the rules of logic are canonical statements of second intentions. What precisely does this mean? The point is simply this: to know what a second intention is is to know *how* to order an object possessed of this second intention in logical or rational discourse. The rules governing such discourse are simply statements in the form of rules or canons of the natures of the different second intentions. For example, to know *what* a definition is is to know *how* to define a term. To know *what* a syllogism is is to know *how* to proceed validly by way of deduction. To know *what* a demonstration is is to know *how* to prove a conclusion. In the act of discourse the intellect orders its objects in accord with the relationships which exist between them. These relationships must be understood or the ordering may well be erroneous. Without an understanding of these relationships their use in rational discourse will certainly not be defensible.

Logic can be characterized as a real science but not as a science of real things. It is a real science in the sense that it is genuinely a science. It is genuinely a science because it resolves its conclusions into self-evident premises in its own order. The particular rules of logic are not arbitrary. They are not presumed without proof. They are not self-evident. Rather, as we have seen, they are proved from self-evident propositions. From a practical point of view, this means that the student is not a logician simply because he knows what the rules of logic are. These, presumably, could be accepted on the authority of a professor or a textbook and merely memorized. The student becomes a logician only when he comes to know what the rules of logic are, and why. He is a logician only in so far as he sees that they cannot not-be what they are by showing that they follow from absolutely incontrovertible first principles of logical procedure.

We say that logic is not a science of real things even though it is a real science. It falls short of being a science of real things because its object is a being-of-the-reason and not a real being. The second intention, which is the object of logic, is a logical relationship belonging to an object *precisely as known*. The property of being tired, or old, or hungry can belong to human nature in the real order. The property of being subject, or middle term, or species can belong to *man* only in so far as *man* exists as an object in the mind. Such properties exist only as conferred by the mind upon objects precisely as objects existing in the mind. Thus, these properties are not, in the technical sense, real. They are beings-of-the-reason. It is impossible to think of them existing anywhere except in the mind. Yet they do really exist in the mind. They are not simply nonentities. Further, they do not come into being in any arbitrary fashion. Second intentions are not real, but they are founded upon the real. For example, the logical relation spoken of as predicate may be founded upon the real relationship between accident and substance. In the proposition,

John is white, white is related to John as predicate because in the real order whiteness is an accident inhering in the substance of John.

We have seen that sciences are either speculative or practical. Logic is a speculative science. Science is practical only in so far as it is of value for something *other than* knowledge. This is not the case with logic. The logician studies second intentions precisely so that he may be able to reason well. The end of logic is sound discourse. This is an end strictly within the sphere of knowledge. Thus logic is a speculative science. Yet logic differs from most speculative sciences. Ordinarily, speculative science is of value because a knowledge of the object of the science in question is worth something in itself. Thus, the metaphysician studies being because it is good to know about being. However, the logician seeks knowledge of an object not for the sake of knowledge of *that* object, but rather for the sake of knowledge of other objects. The logician does not study second intentions because it is good to know about second intentions. He studies second intentions because a knowledge of second intentions will enable him to proceed scientifically in the disciplines which seek knowledge of real things. Logic is not an end in itself. It is nothing if it is not useful. But its use is within knowledge itself, and it remains, then, knowledge for the sake of knowledge. It is at the same time useful and speculative. Perhaps the easiest way to see what we mean by this is to compare logic with metaphysics on the one hand and medicine on the other. The object of metaphysics is being, and its end is knowledge of being. It is speculative without being useful. The object of medicine is the curable, and its end is the curing of the curable. Medicine is practical and useful. The object of logic is the second intention. Its end is knowledge — not of the second intention — but of man in psychology, being in metaphysics, the continuum in geometry, etc. Thus, logic is speculative, but useful.

V. The Logic of the First Operation, the Logic of the Second Operation, and the Logic of the Third Operation

The object of the science of logic is the second intention. The second intention is a relation of the reason which belongs to an object precisely as known. Some second intentions accrue to objects on the level of simple apprehension; others on the level of judgment; and still others on the level of reasoning itself. The end of logic is totally sound discourse, with the emphasis upon the reasoning process. The intellect cannot proceed in an orderly fashion on the level of reasoning so long as it is not well ordered on the level of judgment. Nor can the intellect escape disorder on the level of judgment without proceeding in an orderly fashion on the level of simple apprehension. Thus, there is a natural order to the logician's concern for second intentions. First, he must study those which belong to objects on the level of the first operation, with an eye to the second and third. Then he must study those second intentions which objects acquire on the level of the second operation, with an eye to the third. Only then is he able adequately to study those second intentions which arise on the level of reasoning itself. The science of logic, then, is divided into the logic of the first operation, the logic of the second operation, and the logic of the third operation.

VI. Formal and Material Logic

We have seen that logic is simultaneously the art of sound discourse and the science of second intentions. Certain second intentions are required for the integrity of discourse from the point of view of its form. These are second intentions accruing to natures known precisely in virtue of their logical "position" (i.e., their mode of signification) in discourse. Scientific knowledge of these second intentions is adequate to the art of valid or consistent reasoning — but only to the art of valid reasoning. Other second intentions are required for the integrity of discourse from the point of view of its matter. These second intentions are acquired by natures known precisely in virtue of their intelligible content. The logician who knows these second intentions scientifically is able to determine not only the validity of his discourse but the force of his discourse in reference to the truth and necessity of its conclusions. The first type of second intention specifies that branch of logic spoken of as formal logic. The second type specifies that branch of logic called material logic.

In order to see the difference between a second intention in formal logic and a second intention in material logic let us consider the two propositions *Every man is rational* and *Every man is able to speak*. *Rational* and *able to speak* are both related to *man* as predicate, because each enjoys the same "position" within the formal pattern of thought. But *rational* is an essential predicate related to *man* as specific difference, and *able to speak* is a nonessential predicate related to *man* as a specific property. This difference stems from the intelligible content of the concepts *rational* and *able to speak*. The property of being a predicate is a second intention in formal logic. The property of being a specific difference and the property of being a predicable property are second intentions in material logic. If we refer to the theory of argument we can recall that it is necessary to know what a predicate is simply to be able to reason validly. Yet one could recognize a valid argument without knowing the difference between a specific difference and a property. However, knowledge of the difference between a difference and a property is indispensable for anyone who would try to distinguish between an explanatory proof and an a posteriori proof. The one second intention is of formal import, while the other two are of material import.

The distinction between formal and material logic is most evident (and of prime importance) on the level of the third operation. However, we can distinguish between second intentions of both formal and material import in each of the three operations of the mind. Throughout this book we have treated both types of second intentions as we have progressed successively through the logic of the first operation, the logic of the second operation, and the logic of the third operation. The following schema helps to illustrate the point by recalling one or two examples of both types of second intentions on each level of intellection:

	FORMAL IMPORT	MATERIAL IMPORT
First operation	term universal	genus definition
Second operation	proposition converse	immediate commensurately universal
Third operation	conclusion syllogism	explanatory demonstration dialectical argumentation

The names "material logic" and "formal logic" inevitably cause some difficulty. In a sense both branches of logic are formal and neither material. Both are specified by a type of logical form or second intention. Material logic, no less then formal logic, studies logical relationships which accrue to concepts. Neither studies the intelligible content of the concept itself. This is left, of course, to the other sciences. The relationship of predicate which the concept *man* acquires in the mind when used in propositions is studied by the formal logician. The relationship of species which the concept *man* acquires in the mind with regard to other objects is studied by the material logician. But the nature of man is studied by the psychologist. The second intentions in material logic depend proximately upon the intelligible content of the objects to which they accrue. But material logic, no less than formal logic, presumes to make no judgment precisely of the intelligible content of any logical discourse. The material logician knows what the logical conditions which ensure a certainly true conclusion must be. But the material logician, as material logician, never says that this or that proposition is certainly true or not. Logic is related to each of the other sciences as a tool. It is used by each of them. But it does not do the work precisely of any one of them. This is as true of material logic as it is of formal logic.

The division of logic into formal and material logic admits of a subdivision. Some discourse is so disposed materially as to generate a certainly true conclusion. Other discourse generates only a probably true conclusion. Still other discourse generates merely suspicion. A final type is proportioned simply to the generation of an intellectual inclination. These four are respectively: demonstration, dialectical argumentation, rhetorical discourse, and poetry. The distinction between these four types of discourse is the reason for a division of material logic on the level of the third operation into demonstrative logic, dialectics, rhetoric, and poetics.

VII. Logic as a Common Mode

Logic is a science at the same time both speculative and useful. It is speculative because its end is knowledge (not action), and it is useful because the knowledge it seeks ultimately as an end is not a knowledge of its own object but rather of the objects of the other sciences. Because of this, logic is sometimes called the tool of the other sciences. However, it is more properly thought of as the tool of the intellect or reason in its work in the other sciences. Each science has its own object, but the same reason or intellect is operative in every scientific area. Consequently, there is one general mode of procedure demanded by the nature of the reason in its confrontation of real things. This general method of procedure is logic. Each different science has its own special methodology, a methodology properly suited to the scientific character of its formal object. Each of these special methodologies is a particular contraction of the common method. Logic is analogically common to each different type of scientific methodology. The method of mathematics is different from the method of psychology; the method of psychology is different from the method of metaphysics; and so on. Yet each of these methods is logical. The mathematical methodology is logic in the mathematical mode; the psychological methodology is logic in the psychological mode; and the metaphysical methodology is logic in the metaphysical mode. Each particular methodology is logical. But each is *differently* logical, because each is logical *in its own way*.

This is an extremely important distinction. It is a distinction frequently missed. One difficulty for the student arises from the fact that students must rely heavily on examples to help them to understand logical theory. It must be kept in mind that each example is a particular illustration of logic-employed-in-a-special-field. Thus each example suggested for the purposes of logic might also be suggested as an example of some special scientific methodology. For the purposes precisely of logic the student must slough off on the special character of the example which comes from the special character of the scientific subject matter of the discipline from which the example is taken. He must concern himself only with what is analogically common to this example and similar examples from other scientific areas. It is impossible to suggest an example simply of *being logical*. An example of *being logical* must always be an example of being logical *in mathematics*, or *in psychology*, or *in metaphysics*, or the like. The student must be on his guard lest any example of being logical in this or that field be taken as an example of being logical *simply*.

This last point highlights the useful character of logic. On its own, logic is worth nothing. It is of value only as it looks beyond itself to the other disciplines. There is no value in simply being logical — in fact this is an absurd impossibility. But it is vitally important that in mathematics one be logical *mathematically*, and in psychology that one be logical *psychologically*, and so on in each of the other disciplines.

VIII. Doctrinal Logic and Logic in Use

The fact that logic is a useful science is the basis for the distinction between doctrinal logic (*logica docens*) and logic in use (*logica utens*). Logic, considered simply as the intellectual discipline which is the scientific grasp of second intentions, can be spoken of as doctrinal logic. Doctrinal logic, as we have seen, looks beyond itself to the other sciences for its end. Thus, we can say that doctrinal logic is virtually useful. It becomes actually useful when it is put to the use of the reason. Logic is put to use in one sense whenever it directs the intellect in rational discourse. This can happen in either one of two ways. First of all, logic can direct the intellect in demonstrative discourse, which generates certainly true conclusions by a rigorous resolution to self-evident propositions. Second, logic can direct the intellect in dialectical discourse, the tentative discourse productive of a probable conclusion. Logic is put to use in another sense whenever, as a common science, it supplies a particular science with propositions that can be used as premises leading to a conclusion concerning the subject proper to that particular science. An example of this last use of logic which we have seen before occurs when the psychologist argues from the logical proposition that contraries are in the same genus to the conclusion that the contrary passions of love and hatred flow from the same appetite.

Logic in use, then, has three meanings: namely, the direction of the reason in demonstrations proper to the different sciences, the direction of the reason in different areas of dialectical discourse, and the presentation of common principles which can serve as premises to manifest conclusions in the particular sciences. We can speak of the first two uses of logic as methodological. The third, despite the inevitable terminological confusion, might be spoken of as a doctrinal use (*logica utens ut docens*). In speaking of logic throughout this chapter as a useful science we have been primarily concerned with the first two uses of logic rather than the third.

IX. The Question of Symbolic Logic

It is difficult to find a name which adequately describes logic as we have presented it throughout this book. Inasmuch as the logical works of Aristotle have been the most significant of the primary texts which have contributed to it, and inasmuch as it owes a great deal to the works of men who rank as commentators on Aristotle, it can be spoken of as Aristotelian logic. Custom would allow, too, that it be spoken of as traditional logic. Some would speak of it as Scholastic logic.

There is another discipline, besides that of Aristotelian, Scholastic, or traditional logic, which speaks of itself as a logic. This is, of course, the mathematical or symbolic logic of many modern mathematicians and logicians. Some have claimed that this is an extension and perfection of Aristotelian logic. Some say that it is a universal logic, which either replaces Aristotelian logic or is related to Aristotelian logic as whole to part. Some claim that it is a particular logic which plays the role of instrument in areas foreign to Aristotelian logic. In any event it consists in a complicated system of symbols which are intended to represent *strictly formal* relations between signs. Extremely complicated manipulations of symbols and combinations of symbols are effected with relative ease and admitted *lack of concern for any intelligible content or meaning behind the symbols*. The results are sometimes rather startling in the face of some of the notions enunciated in our study of what can be called traditional logic. We have already noted something of this in our chapter on hypothetical propositions. For the symbolic logician all hypothetical propositions are truth-functional, even the conditionals. For him a true proposition is implied by any proposition, and a false proposition implies any proposition. This, of course, contradicts the doctrine of the conditional proposition as presented in the framework of Aristotelian logic. In the light of this difficulty, and a host of others we shall not go into, what are we to say of the relation between Aristotelian and symbolic logic?

As a matter of fact Aristotelian and symbolic logic are not univocally logics. Neither is an extension of the other, nor is either related to the other as whole to part. Aristotelian logic, as we have seen, is an instrument of the intellect considered precisely as an ability to know the intelligible; i.e., to grasp the meaningful. Logic, thus considered, cannot be *purely* formal in the sense in which the symbolic logician conceives of his logic. It is true that logical relations can sometimes be more easily considered when represented by symbols. As a matter of fact we have made liberal use of symbols throughout this book. But the relationships we have symbolized — even in formal logic — have a real foundation. We can never mistake the symbol for the relationship and never prescind from the real foundation that specifies the relationship — not even in formal logic. The symbolic logician does this. The relationships he is concerned with are not second intentions. They are not founded upon the real. At best they are strictly formal relationships between symbols which need not stand for anything. The formal object of the science of Aristotelian logic is the second intention. Whatever the formal object of symbolic logic is, it is not the second intention. Since sciences are specified by their formal objects, these two sciences cannot be identified in any way. If one is logic, the other is called logic by analogy at best. The use of symbols is highly desirable in logic, even Aristotelian logic, but so-called symbolic logic is not logic in the sense in which Aristotelian logic is logic. Whatever it is, and whatever its use, symbolic logic is not an extension of Aristotelian logic, nor is Aristotelian logic an extension of symbolic logic. Since symbolic logic is not an instrument of the intellect *as it confronts the real*, it should not be considered to be a rival of Aristotelian logic, whose avowed end is finally scientific knowledge of real things in the sciences which are concerned with real things. This is not to deny mathematical or symbolic logic a place in the academic sun. In its own right, and in reference to its proper end, it may be a very significant discipline. The point here is simply that Aristotelian logic and symbolic logic are radically distinct disciplines.

THE NATURE OF LOGIC

Exercise XXII

1. Define: art, science, liberal art, servile art, speculative science, practical science, useful science, second intention, first intention, formal logic, material logic, demonstrative logic, dialectics, rhetoric, poetics, doctrinal logic, logic in use, symbolic logic.
2. What does it mean to say that it is not the business of the logician to investigate the nature of logic?
3. Compare the nature of art with the nature of science.
4. What is the difference between a servile art and a liberal art?
5. What is the difference between a speculative science and a practical science?
6. Why is logic said to be an art? Why is it a liberal art?
7. What does art demand on the part of the matter of the artistic work? What does art demand on the part of the artist? Apply this to the special case of logic.
8. Why is logic said to be a science?
9. Why must logic be a science in order to be the art it is?
10. What does it mean to say that logic is a real science but not a science of the real?
11. What does it mean to say that the rules of logic are canonical statements of the natures of second intentions?
12. How can the science of logic be speculative and at the same time useful? Compare logic with medicine on the one hand and metaphysics on the other in terms of object and end.
13. How can logic be simultaneously a science and an instrument for science?
14. Explain the division of logic into the logic of the first operation, the logic of the second operation, and the logic of the third operation.
15. Explain the division of logic into formal and material logic. Distinguish between these in terms of object and end.
16. In what sense are both formal logic and material logic formal? In what sense is neither *purely* formal?
17. What is the rationale of the division of material logic into demonstrative logic, dialectics, rhetoric, and poetics?
18. Why is logic more properly said to be an instrument of the reason rather than an instrument of the sciences?
19. How is the proper methodology of a particular science related to logic?
20. What is the difference between doctrinal logic and logic in use?
21. What are the three uses of logic? Which have we been most concerned with?
22. Aristotelian logic makes use of symbols, without being symbolic logic. Explain.

Index

Absolute supposition, 109, 116
Abstract concept, 35 n
Abstraction, from both individuating characteristics and the act of existing, in simple apprehension, 19 f, 29; of the intelligible from the sensible, 4 f, 19; twofold, by way of apprehension and by way of negative judgment, 19, 37 f
Abstractive induction, 281 f
Accent, fallacy of, 299 f
Accident, 56; categorical, see Accident (categorical); fallacy of, 300 f; predicable, see Accident (predicable)
Accident (categorical), 59 ff, 62 ff
Accident (predicable), 49 f; nature of, 42 f; wholly inseparable, 50; wholly separable, 50
Action, category of, 59 ff, 63
Adversative proposition, 132 f
Albert, St., primary source in logic, ix
Alternative exclusive syllogism, see Exclusive alternative syllogism
Alternative inclusive syllogism, see Inclusive alternative syllogism
Alternative proposition, 135 ff; exclusive, 136 f; inclusive, 135 f; reduced to conditional, 138
Ambiguity, fallacy of, 298
Analogous terms, presenting special difficulty in discourse, 115; unsuited to the categories, 58
Analogy, 106; of names, 55
Antecedent, of conditional proposition, 133; and consequent of the argumentation, 175
Antepredicaments, purpose of, 55 ff
A posteriori demonstration, 265 ff; cannot be explanatory, 265 ff; of the fact, two types, 268 f
A priori demonstration, 265 ff; explanatory, 265 ff; of the fact, two types, 267 f
Argument, see Argumentation
Argumentation, antecedent and consequent of, 175; categorical, 179; deductive, 178 f; demonstrative, 179; demonstrative, see also Demonstration; dialectical, 179; dialectical, see also Dialectic; and its elements, 174 f; expression of object of reasoning, 22, 24; fallacious, 179, 297 ff; hypothetical, 179; inductive, 178 f; matter of, 175; sophistical, 179; types of, 178 ff
Argumentum ad baculum, 302 f
Argumentum ad hominem, 302
Argumentum ad misericordiam, 303
Argumentum ad populum, 302
Argumentum ad verecundiam, 303
Aristotelian logic, 142 n
Aristotle, preliminaries to categories, 55; primary source in logic, ix; on self-evident propositions, 285; sorites, 248 ff; use of term "enthymeme," 244
Art, liberal or servile, 311; nature of, 310 f
Axiom, necessary for any science, 285

Begging the question, 303 f
Being, analogically divided into categories, 61 f; finite, can be categorized, 58; infinite, cannot be categorized, 58; as quasi-genus, 62; real, 56 f; real, actual or possible, 99; real, can be categorized, 57; of the reason, 56 f, 99; of the reason, cannot be categorized, 57
Bourke, V., 221

Cajetan, Thomas de Vio, primary source in logic, ix
Categorical proposition, 120 ff, 129; expressed as hypothetical, 139
Categorical syllogism, basic principles for, 189 f; defined, 182; elements of, 183 f; expressed as hypothetical, 227 f; figures of, 185 f; general rules for, 196 ff; general rules for, in reference to basic principles for, 196; moods of, 188 f; reduction of moods to first figure, 190 f; reduction of moods to first figure, direct and indirect reduction, 213 ff; reduction of moods to first figure, directions for each, 216; special rules for each figure, 205 ff; valid moods, 208 ff
Categories (predicaments) 25; division into ten, 59 ff; division of the universal essentially taken, 39, 41; nature and divisions of, 54 ff; as related to the predicables, 65 f; as ultimate genera, 62;

325

univocally divided into species, 61; what kind of being can be put in, 57 ff
Causal definition, 77 ff
Causal proposition, 132 f
Cause, efficient, 76; extrinsic, 76; final, 76; formal, 76; four, 75 ff; intrinsic, 76; material, 76; role of, in explanatory demonstration, 266 f
Certitude, 11, 179, 258 ff, 319
Codivision, 89
Collective supposition, 112, 117
Commensurate universality (convertibility), required for strict demonstration, 264
Complex question, fallacy of, 305
Composite expression, 101 f
Composition and division, fallacy of, 299
Compound proposition, 120, 129 ff; more or less complex, 129
Comprehension, of a concept, 25 ff; measured in terms of infinite multitude, 29 f; priority over extension, 30; of terms, in propositions, 125
Comprehensive notes, intelligibly but not physically distinct, 26; known explicitly only with difficulty, 27; in less strict sense, 28; not parts on a par with one another, 27
Concept, abstract, 35 n; comprehension of, 25 ff; concrete, 35 n; extension of, 25; formal, 22; objective, 23
Conclusion, in categorical syllogism, 183 f; direct, 184 f; of hypothetical syllogism, 223; indirect, 184 f; a new truth potentially in its premises, 174
Concrete concept, 35 n
Conditional proposition, 133 ff; reciprocal, 134 f; simple, 133 f
Conjunctive proposition, 129 ff
Consequent, and antecedent of the argumentation, 175; of conditional proposition, 133; fallacy of, 305
Contradiction, between concepts, 66 f; between propositions, 148 f; of singular propositions, 163
Contradictory terms, as used in obversion, 160
Contraposition, between propositions, importance of, 162; of propositions, 161 ff; questionable for singular propositions, 166
Contrariety, between concepts, 68; between propositions, 149 f; impossible for singular propositions, 163
Contrary terms, as used in obversion, 160

Converse, 155 ff
Conversion, accidental, of singular propositions, 165; of propositions, 155 ff; of propositions, accidental, 156 ff; of propositions, importance of, 156 f; of propositions, simple, 155 ff; of propositions, when legitimate, 157 f; simple, of singular propositions, 165
Convertend, 155 ff
Convertibility (commensurate universality), required for strict demonstration, 264
Copula, of the categorical proposition, 120; of the compound proposition, 120, 129; establishing a suppositional demand for the subject, 114 f; of proposition, form of verb "to be," 102
Copulative proposition, 130 ff; occultly, 130 f; openly, 130; plus, 133

Deduction, 178 f; insufficiency of, 280 f
Definition, 67 ff; of an accident, 81; of an analogical concept, 81; causal, 77 ff; of characteristically collective terms, 113; of characteristically divisive terms, 113; discloses the comprehension of a concept, 28; essential, 77 ff; essential, two modes, 78 f; by extrinsic cause, 77 ff; frequently difficult of attainment, 91; by intrinsic causes, 77 f; modes of, in reference to abstraction from matter, 270 f; nominal, 73 ff; nonessential, 77 ff; by property, 77 ff; real, 73 ff; real, types of, 77 ff; rules for, 80 ff; of an ultimate genus, 81; value of, along with division, 90 f; by way of combination of accidents, 80; by way of etymology, 74 f
Demonstration, 179, 258 ff; an analogous notion, 259; a posteriori, 265 ff; a priori, 265 ff; difficult to achieve, 271; explanatory (propter quid), 264 ff; explanatory, three types of, 265 f; of the fact (quia), 264 ff; from necessary premises to a necessary conclusion, 262 f; premises of, 261 f; prescientific knowledge required for, 260 f; role of causes in, 286 f; as the scientific syllogism, 259; in strictest form, 259 f, 265 f; terminated in the real, 273; types of, 264 ff; weaker forms of each type, 269; yields certainly true conclusions, 179
Demonstrative logic, 319
Deprecative statement, 102
Dialectic, 179, 285 ff, 319; leads to prob-

able conclusion, 179, 271 ff; as preparation for demonstration, 275 f; three purposes of, 276; unterminated, i.e., remains in the reason, 273 f
Dialectical argumentation, see Dialectic
Dichotomous division, 86
Dictum de nullo, 189 f
Dictum de omni, 189 f
Difference, defined, 46 f; generic, 46; predicable, nature of, 42 f; specific, 46
Dilemma, 251 ff; complex constructive, 252 f; complex destructive, 253; simple constructive, 252; simple destructive, 252
Discourse, logical, in more and less strict sense, 84; most strictly realized in reasoning, 176; rational, 7; in strict and less strict senses, 73
Disjunctive proposition, 137; reduced to conditional, 138
Disjunctive syllogism, with multimembered major, 238 f; special rules, 234 ff
Division, logical, 72, 84 ff; logical, of analogical wholes, 85 f; logical, dichotomous and nondichotomous, 86; logical essential, 85; logical nonessential, 85; physical, 87 ff; physical, variety of, 87 f; rules for, 88 f; for the sake of definition, 90 f; value of, along with definition, 90 f
Divisive supposition, 112

Efficient cause, 76
Enthymeme, a rhetorical syllogism, 244
Enumeration, inductive, complete, always sufficient, 293; inductive, complete, only of species and not of singulars, 292 f; inductive, must be discriminative to be sufficient, 291 f; inductive, ordinarily generates probability at best, 293 f; inductive, sufficiency of, 290; rather than division, 88; of singulars, principle for mediate induction, 288 ff
Epicheirema, 246 ff
Equivocal terms, invalidating discourse, 115; unsuited to the categories, 58
Equivocation, fallacy of, 298
Equivocity, 106; of names, 55
Essence, both substantial and accidental, 66; object of simple apprehension, 6, 17, 19, 97
Essential definition, 77 f; two modes, 78 f
Essential division, 85
Evidence, intrinsic or extrinsic, 100 f
Exceptive copulative proposition, 131

Exclusive alternative syllogism, special rules of, 233; with multimembered major, 238
Exclusive copulative proposition, 131
Existence, actual or possible intended in propositions, 117; admitting of different modes and grounding different judgments, 98 f; known along with essence in the judgment, 97; object of judgment, not simple apprehension, 19; possible or actual intended in propositions, 113 f
Explanatory demonstration, 264 ff; three types of, 265 f
Exponible proposition, 130
Extension, of a concept, 25, 29 ff; inversely proportional to comprehension, 30; of terms in propositions, 110, 125

Faith, assent on extrinsic evidence, 100 f
Fallacy, of accent, 299 f; of accident, 300 f; of ambiguity, 298; of arguing beside the point, 301 ff; argumentum ad baculum, 302; argumentum ad hominem, 302; argumentum ad misericordiam, 303; argumentum ad populum, 302; argumentum ad verecundiam, 303; of begging the question, 303 f; beyond language, 300 ff; of complex question, 305; of composition and division, 299; of the consequent, 305; of equivocation, 298; of false cause, 304 f; of language, 298 ff; of mistaking qualified for absolute, 301; of similar word construction, 300; of vicious circle, 304
False cause, fallacy of, 304 f
Falsity, and truth of propositions, 101
Figures, of categorical syllogism, 185 f; of categorical syllogism, first and most perfect, 187; of categorical syllogism, no fourth, 187 f; of categorical syllogism, three and only three, 185 f; of hypothetical syllogism, 223 f
Final cause, 76
Finite being, can be categorized, 58
Formal cause, 76
Formal concept, mental expression of object known in simple apprehension, 22; various names for, 24
Formal logic, 317 ff; concerned with validity, 9 ff
Formal object, 4
Formal supposition, 107 ff; types of, 108 f

Genus, defined, 44 ff; predicable, nature

of, 42 f; proximate, 45; remote, 45; ultimate, 45
Glanville, John J., 266 n
Grammatical expression, not always identical to logical expression, 123
Grammatical usage, of terms, not to be confused with logical usage, 111

Having possession, category of, 59 ff, 64
Hollenhorst, G. Donald, 266 n
Houde, Roland, 277 n
Hypothetical proposition, 129 f; fundamentally conditional, 137 f
Hypothetical syllogism, 222 ff; with compound components, 239 f; conclusion of, 223; figures of, 223 f; major premise, 223; minor premise, 223; moods of, 223 f; with multimembered major premise, 237 ff; types of, 223; see also Disjunctive syllogism, Exclusive alternative syllogism, Inclusive alternative syllogism, Reciprocal conditional syllogism, Simple conditional syllogism

Idea, see Formal concept
Illicit major term, 198 f
Illicit minor term, 198 f
Immediate induction, of self-evident propositions, 281 f
Imperative statement, 102
Implication, formal, 142 n; material, 142 n
Inclusive alternative syllogism, special rules, 231 f; with multimembered major, 237 f
Induction, 178 f; abstractive, 281 f; immediate, of self-evident propositions, 281 f; mediate, 282, 287 ff; mediate, compared with deduction, 289 f; mediate, compared with induction of self-evident proposition, 288 f; Mill's canons for, 295; necessary for deduction, 281; relation to deduction, 290; three types of, 281 f
Inference, immediate impossible in strict sense, 180; strictly taken, limited to reasoning, 147
Inferior, subjective part, 31 ff
Infinite being, cannot be categorized, 58
Interrogative statement, 102

John of St. Thomas, primary source in logic, ix
Judgment, 97 ff; attributive, 97; certain or probable, 100; complexity of its object, 17; merely existential, 97; and the motive for assent, 99 ff; the operation of composition or division, 98; pre-eminence as an intellectual operation, 99; the second operation of the intellect, 5 f

Knowledge, effected through formal signs, 21 f; intellectual, 2 ff; sensory, 2 ff

Language, fallacy of, 298 f
Larkin, Vincent R., 277 n
Logic, Aristotelian, 142 n; art of, 2; as "art of arts," 2; art of rational discourse, 7; as common mode to be contracted to special methodologies, 319 f; demonstrative science of logical rules, 313; divided into branches by three operations of intellect, 316 f; divisions of, according to operations of intellect, also into formal and material, 9 ff; doctrinal and in use, 320 f; formal and material, 317 ff (see also Formal logic, Material logic); as liberal art, 312; mathematical, 142 n, 317 ff, 321 f; natural, 1 f, 5; nature of, 309 ff; perfection of the intellect not the senses, 5; as "rational science," 2; real science, but not a science of the real, 315 f; science of second intentions, 7; as speculative but useful science, 312 ff, 316; studies logical forms and not intelligible content of objects, 318 f; symbolic, 142 n, 317 ff, 321 f; in use, three uses for, 320 f
Logical relations, the subject matter of logic, 7 ff
Logical supposition, 108, 116
Logico-real supposition, 109 f, 116

Major premise, in categorical syllogism, 183 f; of hypothetical syllogism, 223
Major term, illicit, 198 f
Material cause, 76
Material logic, 317 ff; branches of, 319; concerned with probative force, 11 f
Material object, 4
Material supposition, 107 f, 115 f
Mathematical logic, 142 n, 317 ff, 321 f; no rival of Aristotelian, 322
Matter, proximate and remote, of the argumentation, 175
Mediate induction, 282
Mental term, see Formal concept
Mental word, see Formal concept
Methodology, scientific, 319 f

INDEX

Middle term, in categorical syllogism, 183 f; in demonstration, 270; example of a logical relation, 8; unextended, 199 f
Mill, John Stuart, canons of induction, 295; criticism of syllogism, 191 f
Minor premise, of categorical syllogism, 183; of hypothetical syllogism, 223
Minor term, in categorical syllogism, 183 f; illicit, 198 f
Modal proposition, 14
Moods, of categorical syllogism, 188 f; of categorical syllogism, indirect in second and third figures, 211 ff; of hypothetical syllogism, 223 f; valid, of categorical syllogism, 208 ff

Natural logic, 1 f, 5
Nature, object of simple apprehension, 17, 23; taken absolutely, 36 f; see also Essence
Natures, complete, can be categorized, 58; complex, cannot be categorized, 57; incomplete, categorized only reductively, 58; incomplete, can be categorized, 57
Nominal definition, 73 ff
Nominalists, notion of extension, 30
Nondichotomous division, 86
Nonessential definition, 77 ff
Nonessential division, 85
Noun, element of proposition, 102

Object, formal, 4; material, 4
Objective concept, the object known in simple apprehension, 23
Obverse, 159 f
Obversion, of propositions, 159 ff; of propositions, importance of, 160; of singular propositions, 165 f
Obvertend, 159 f
Operation, immanent, 63; transitive, 63
Opposition, between concepts, 66 ff; between concepts, modes of, in reference to division, 85; between propositions, 148 ff; square of, 154

Particular supposition, 110 ff, 117
Passion, category of, 59 ff, 64
Perseity, four modes of, 263 f; not to be confused with self-evidence, 284 n; required for strict demonstration, 263 f
Personal supposition, 109 f, 112
Physical division, 87 ff

Plural, of grammatical concern, not logical, 123
Poetic argumentation, see Poetics
Poetics, 272, 319
Polysyllogism, 247 f
Porphyry, Neoplatonic commentator on Aristotle, 51; primary source in logic, ix; Tree of, 51
Position, category of, 59 ff, 64
Predicables, 41 ff; distinguished from categories, 54; a division of only some predicates, 50 f; division of the universal, 39; division of the universal denominatively taken, 41 ff; essential, 47; essential and nonessential, 42 f; as related to the categories, 65 f
Predicaments, see Categories
Predicate, 97 ff; establishing a suppositional demand for the subject, 114 f; extension of, 124 f; what it is, 102 f
Predication, essential and nonessential or denominative, 33
Premises, for demonstration, 272; for dialectical argumentation, 272 f; major, in categorical syllogism, 183 f; major of hypothetical syllogism, 223; major, see also Major premise; minor, in categorical syllogism, 183 f; minor, of hypothetical syllogism, 223; minor, see also Minor premise
Principle, of contradiction, a methodological axiom, 285; of the separating third, 189 f; of sufficient enumeration, 290; of triple identity, 189 f, 289
Privation, a mode of opposition between concepts, 67
Privative terms, as used in obversion, 160
Probability, 11
Probative force, the concern of material logic, 11 f
Property, defined, 47 ff; generic, 48; predicable, nature of, 42 f; relation to difference, 48; specific, 48
Proposition, adversative, 132 f; alternative, 135 ff; alternative exclusive, 136 f; alternative inclusive, 135 f; based on intrinsic or extrinsic evidence, 100 f; categorical, 129; causal, 132 f; compound, 120, 129 ff; compound, more or less complex, 129; compound, symbols for, 139 f; conditional, 133 ff; conditional, reciprocal, 134 f; conditional, simple, 133 f; conjunctive, 129 ff; contraposition of, 161 ff; conversion of, 155 ff; copulative, 130 ff; copulative, occultly,

130 f; copulative, occultly, exceptive, 131; copulative, occultly, exclusive, 131; copulative, occultly, reduplicative, 131; copulative, openly, 130; copulative plus, 133; disjunctive, 137; disjunctive, not simply contradiction of copulative, 140; divided according to both quantity and quality, 126 f; division of, 143; elements of, 120; exponible, 130; expression of object of judgment, 22, 24; factually evident, 100 f; having force of singular with nonpersonal subjects, 124; hypothetical, 129 f; hypothetical, fundamentally conditional, 137 f; hypothetical, truth functionally taken, 142 f n; modal, 121; modal, with mode of contingency, 121; modal, with mode of impossibility, 121; modal, with mode of necessity, 121; modal, with mode of possibility, 121; obversion, 159 ff; quality of, affirmative or negative, 122 f; quantity of, universal, particular, or singular, 122 f; related, mutually indices for truth-status of one another, 147; self-evident, 100 f; self-evident, the absolute premise for demonstration, 262 f; self-evident, basic truth of the scientific syllogism, 283; self-evident, immediate, 282 f; self-evident, indemonstrable, 283; self-evident, induction of, 281 f, 286 f; self-evident, principle of demonstration, 282 ff; self-evident, types of, 284 ff; simple or categorical, 120 ff; simply attributive, 121; strict logical form of, 102 f; truth-functional, 141 ff; types of, 120 ff

Quality, category of, 59 ff, 62 ff; species of, 63
Quantity, category of, 59 ff, 62 f
Quiddity, see Essence

Real being, actual or possible, 99
Real definition, 73 ff; types of, 77 ff
Real supposition, 108, 116; types of, 109 f
Reasoning, nature of, 173 ff; the need for, 173 f; for the sake of judgment, 173 f; the third operation of the intellect, 5 ff
Reciprocal conditional syllogism, special rule, 229 f
Reduplicative copulative proposition, 131
Relation, belonging to singular propositions, 163 ff; between propositions, 147 ff; between propositions, complex, 161 f; a category, 59 ff, 63; logical, second intention, 313 f; mode of opposition between concepts, 68
Rhetoric, 319
Rhetorical argumentation, 272

Science, 258 f; empiriological, 274 f; nature of, 310 f; speculative, divided by different modes of defining, 270 f; speculative or practical, 311
Scientific method, of empiriological science, an instance of dialectic, 274 f
Second intention, 312; canonically expressed in logical rules, 314; logical relations, 7; nature of, 313 ff; object of science of logic, 314 f
Sequential nexus, in hypotheticals, 130
Sign, formal or instrumental, 21 ff; import in logic, 20 ff; natural or conventional, 20 f; nature and divisions of, 20 ff
Signification, and supposition, 106 f; of terms, 106, 115
Simon, Yves R., 266 n
Simple apprehension, the first operation of the intellect, 5 f; incomplexity of its object as compared to object of judgment, 17, 55; its object simple or complex, 17 f; its object, simple or complex, in reference to categories, 55 f; nature of, 17 ff; a truncated act of knowing, 97
Simple conditional syllogism, 225 ff; rules for, 226
Simply attributive propositions, 121
Singular, as characterizing the sense object, 4; as opposed to universal, 56
Singularity, of things in the real, 36 f
Singular supposition, 110 ff, 117
Sophistical argumentation, 129
Sophistry, 297
Sorites, 248 f; Aristotelian, 248 ff; Goclenian, 249 ff
Species, from point of view of subjectibility, 46; one per subject, 44; predicable, nature of, 42 ff
Subalternand (superior), 152
Subalternate (inferior), 152
Subalternation, of propositions, 152 f; of singular propositions, 164 f
Subcontrariety, between propositions, 150 f; of singular propositions, 163 f
Subdivision, 89
Subject, logical, 97 ff, 102 f
Substance, 56; category of, 59 ff, 62

Superior, logical whole or potential whole, 31 ff
Supposition, absolute, 109, 116; absolute, virtually universal, 111, 116; collective, 112, 117; divisive, 112; formal, 107 ff; formal, types of, 108 f; logical, 108, 116; logical, functions as though singular, 112; logico-real, 108 f, 116; logico-real, functions as though singular, 111; material, 107 f, 115 f; particular, 117; particular or undistributed, 110 ff; personal, 109 f, 116; of predicates, usually real and personal, 107; real, 108, 116; real, types of, 109 f; singular, 110 ff, 117; of terms, 106 ff, 115 ff; universal, 117; universal or distributed, 110 ff
Syllogism, 182; abbreviated, 243 ff; abbreviated, grammatically only, 244; categorical, 182 ff, see also Categorical syllogism; categorical, elements of, 183 f; complex in combination, 253 f; complex, 243 ff; disjunctive, 234 ff, see also Disjunctive syllogism; exclusive alternative, 232 ff, see also Exclusive alternative syllogism; "expository," 193 f; hypothetical, 222 ff, see also Hypothetical syllogism; hypothetical, types of, 223; inclusive alternative, 231 f, see also Inclusive alternative syllogism; reciprocal conditional, 229 ff, see also Reciprocal conditional syllogism; reduction of alternative and disjunctive to conditional, 236 f; simple conditional, 225 ff, see also Simple conditional syllogism; with a justification for its premises (epicheirema), 245 ff
Symbolic logic, see Mathematical logic

Terms, expressed indefinitely, 123; expression of object of simple apprehension, 24; major, in categorical syllogism, 183 f, see also Major term; middle, in categorical syllogisms, 183 f; middle, role in demonstration, 270 f, see also Middle term; minor, in categorical syllogism, 183 f, see also Minor term; nonsupposing subject, 114 f; and their signification, 106 ff; and their supposition in propositions, 106 ff; unity of, in logical discourse, 115; unity of, in propositions, 153 f
Thesis, necessary for a given science, 285
Thomas Aquinas, St., 93, 221, 277 n; definition of logic, 2; on kinds of self-evident propositions, 284 f; primary source in logic, ix
Transcendental object, 58
Truth, and falsity of propositions, 101; found properly in judgment, 101; and validity, 176 ff
Truth-functional proposition, 141 ff
Truth table, 141 ff; for hypotheticals strictly taken, 142; for hypotheticals truth functionally taken, 142; limited use for hypotheticals strictly taken, 141 f; for square of opposition, 155

Unextended middle term, 199 f
Unity of terms, required in logically related propositions, 153 f
Universal, as characterizing the intelligible object for man, 4 f; divided into predicables and categories, 39; does not falsify the real, 38; nature of, 56; problem of, 35 ff
Universality, enumerative, 193; of object in the intellect, 36 f; required for strict demonstration, 263; true, rather than enumerative, 193
Universal supposition, 110 ff, 117
Univocal terms, readily suited for discourse, 115; suited to the categories, 58
Univocity, 106; of names, 55

Validity, the concern of formal logic, 9 ff; and truth, 176 ff
Verb, element of proposition, 102; as verb-copula, 102; as verb-predicate, 102
Vocative statement, 102

Wallace, William A., O.P., 278
When, category of, 59 ff, 64
Where, category of, 59 ff, 64
Words, as instrumental conventional signs of meanings, 22 f

www.ingramcontent.com/pod-product-compliance
Lightning Source LLC
Chambersburg PA
CBHW071228230426
43668CB00011B/1355